W9-AQF-196

CALCULUS
A SHORT COURSE

Michael Gemignani

Associate Professor of Mathematics, Smith College

W. B. SAUNDERS COMPANY • PHILADELPHIA • LONDON • TORONTO

W. B. Saunders Company: West Washington Square
Philadelphia, Pa. 19105

12 Dyott Street
London, WC1A 1DB

833 Oxford Street
Toronto 18, Ontario

Calculus: A Short Course ISBN 0-7216-4097-4

 Made in the United States of America. Press of W. B. Saunders Company. Library of Congress catalog card number 76-173333. AMS Subject Classification number 2620.

Print No.: 9 8 7 6 5 4 3 2

To Janet, Joey, Janie, and Phil

Preface

This is a calculus book directed at beginning students of the management and social sciences. Topics range through differential and integral calculus of one variable to introductory calculus of several variables. Applications are discussed wherever possible. However, I have avoided motivating ideas through illustrative examples of applications that require technical information not usually part of a reader's general knowledge. Enough material is provided, with some flexibility, for a one semester course. Background knowledge required of the reader is some high school algebra and a little analytic geometry.

The mathematical ideas presented in the book are not new, but they are usually difficult for beginners. The language barrier alone is formidable. My principle objective, therefore, has been effective communication. I have searched for the best possible means of presenting interesting ideas and useful techniques for the book's intended audience.

To the best of my ability, I have tried to keep the writing free of obstacles to understanding. Paragraphs are short, sentences simple and usually direct. I have sought always to be intuitive and discursive rather than formal and abstract.

Each chapter opens with a brief summary of key ideas presented and explored throughout the chapter. This opening may help warm the often chilling encounter many beginning calculus students have with new mathematical ideas and the language in which they are couched.

The illustrations are an important constituent in the overall structure of each chapter. In the mathematical ideas of analysis, the right picture carefully explained is often the first step toward understanding. The book is full of pictures. The legends accompanying each illustration provide a simplified explanation of the content of the illustration and its relation to the text. Working together, the illustrations and accompanying legends provide a supportive frame for each chapter.

Illustrative examples serve a dual role: to provide motivation of ideas and to provide models for the solutions to some of the exercises. With this in mind, I have tried to provide illustrations that implicitly provide solutions to a fair number of exercises. For convenience to instructor and student, the exercises are organized into three categories: routine, challenging, and theoretical. The exercises are meant to be an integral part of the book. Answers to over half of the problems appear in the appendix.

Each chapter closes with a summary designed for review. A set of supplementary exercises appear after the summary. These exercises are designed to test the student's grasp of the general ideas and techniques explored in the chapter.

In the many revisions and rewritings of the book, I have been guided by many fine critical reviews provided by mathematicians involved in teaching or planning short calculus courses. If the book is indeed a clear and useful guide to the calculus for students of the management and social sciences, then much credit must go to these individuals. Special thanks go to William G. Chinn of San Francisco City College, Miss Debbie Epstein, student of the University of Rochester, George Gastl of the University of Wyoming, Morton Harris of the University of Illinois, Chicago Circle, Donald Kerr of the University of Indiana, Norman Luther of Washington State University, John Pfaltzgraff of the University of North Carolina, and Glenn Pfeifer of the University of Arizona.

Contents

CALCULUS

A SHORT COURSE

Vocabulary

set
intersection
union
absolute value
interval
equation
inequality
line
slope
function
graph

Chapter 1

Sets, Functions and Graphs

1.1 SETS

Mathematics almost invariably concerns some kind of collection: a collection of numbers, instructions for a computer, points in a plane. We therefore begin with a brief discussion of collections in general. The collections or *sets* we will deal with later will be more specialized and exhibit many properties beyond those of a simple collection of objects.

If we are to deal precisely with any collection of items, then the collection itself should be clearly and unambiguously defined. We should be able to decide which items are and which are not members of the collection. If, given a collection S of objects, we can determine whether any one is or is not in S, then we say that S is *well-defined.*

Definition 1: *A* ***set*** *is any well-defined collection of objects. The objects contained in a set are called* ***elements*** *or* ***points*** *of the set..*

Example 1: A deck of computer cards constitutes a set, and each card is an element of the set.

Example 2: The collection of great American novels is not a set, since differences of opinion exist concerning which American novels are "great." The collection of American novels selected as "great" by some particular critic would, however, be defined and, hence, would be a set.

Example 3: The collection of all persons holding accounts at a certain bank is a set. The collection of items manufactured by a particular factory on a given day is also a set.

Various notational devices are used in mathematics to express concisely ideas that would require more space to write out completely. We now introduce some of the notation in general use with regard to sets.

NOTATION: There are various ways of denoting sets. The elements of a set may be listed between braces, $\{\ \ \}$. For example, $\{1, 2, 3\}$ is the set consisting of the elements 1, 2, and 3. A set may also be described by specifying the conditions for the elements. For this method we use a letter symbol to stand for a representative element, and we state the requirements that are to be met by each element. In this notation, the symbol $|$ is used to mean "such that." For example, $\{x \mid x$ is an American citizen$\}$ is the set of all x such that x is an American citizen. We can paraphrase this simply as the set of American citizens. Similarly, $\{w \mid w$ is red and w is a barn$\}$ denotes the set of all red barns.

Given two sets, we can form new sets, the *intersection* and the *union:*

Definition 2: *Let S and T be sets. Then the* **intersection** *of S and T is the set consisting of all elements common to both S and T. We denote the intersection of S and T by* $S \cap T$. *Thus,* $S \cap T = \{x \mid x$ *is an element of S and x is an element of T*$\}$.

The **union** *of S and T is the set consisting of all elements that belong either to S or to T (or both S and T). We denote the union of S and T by* $S \cup T$. *Thus,* $S \cup T = \{y \mid y$ *is an element of S and/or y is an element of T*$\}$.

Example 4: If $S = \{x \mid x$ is red$\}$ and $T = \{y \mid y$ is a house$\}$, then $S \cap T$ is the set of red houses, while $S \cup T$ is the set of all objects that either are red or are houses or are both red and houses.

Example 5: If $S = \{x \mid x$ is a real number and x is less than $5\}$ and $T = \{y \mid y$ is a real number and y is greater than $3\}$, then $S \cap T = \{w \mid w$ is a real number between 3 and $5\}$; while $S \cup T = \{z \mid z$ is a real number$\}$ (since every real number is either greater than 3 or less than 5).

NOTATION: We use $<$ to denote *is less than*, while $\leq$ denotes *is less than or equal to;* $>$ denotes *is greater than*, while $\geq$ denotes *is greater than or equal to*. Thus, $3 < 5$ and $6 > -1$; $\{x \mid 3 < x \leq 5\}$ would denote the set of all real numbers greater than 3 *and* less than or equal to 5.

It is possible that two sets contain no elements in common, in which case their intersection would contain no elements. The set containing no elements is called the empty set, and is generally denoted by $\varnothing$. For example, $\{x \mid x \text{ is a triangle with four sides}\} = \varnothing$.

Definition 3: *A set S is a* ***subset*** *of a set T if each element of S is an element of T (in other words, if the elements of S comprise part or all of the elements of T). We denote "S is a subset of T" by $S \subset T$.*

Two sets S and T are ***equal*** *if the sets contain precisely the same elements. If S equals T, we write $S = T$.*

Thus, the set of odd integers is a subset of the set of all integers, and the set of real numbers less than 1 is a subset of the set of real numbers less than 2.

Two sets will be equal if the criteria for membership in the sets are equivalent; in other words, two sets S and T will be equal if something fulfills the condition for membership in S if and only if it also fulfills the condition for membership in T. This, in turn, is equivalent to saying that $S = T$ if each element of S is also an element of T and each element of T is also an element of S, or

$$S = T \quad \text{if} \quad S \subset T \quad \text{and} \quad T \subset S.$$

Equal sets appear in Examples 6, 8, 9, and 10 below.

Most of the sets considered in our subsequent discussion will be sets of real numbers. Rather than attempt a formal definition of real numbers, we will merely consider the set R of real numbers to consist of all numbers that can be represented as decimals (either ending or unending), as well as all of the usual integers and fractions such as 5, -3, 1/2, and 65/879. Both 134.04 and 0.00345 are examples of ending decimals, while $1/3 = 0.3333\ldots$ is an unending decimal. *Irrational numbers*, real numbers that cannot be written as the quotient of two integers, can always be represented as unending decimals. For example, $\sqrt{2}$ and π are both irrational numbers. In decimal form, 3.14159 represents π correct to 5 decimal places, but 3.14159 can be extended indefinitely to give better and better approximations of π. We can never represent π exactly in decimal form, since the decimal representation of an irrational number involves infinitely many digits with no regular pattern in the occurrence of the digits. Every ending decimal is a rational number; for example, $0.34 = 34/100$ and $1.357 = 1 + 357/1000 = 1357/1000$. An unending decimal may be rational, as in the case of 1/3.

AN ORDERING OF THE REAL NUMBERS

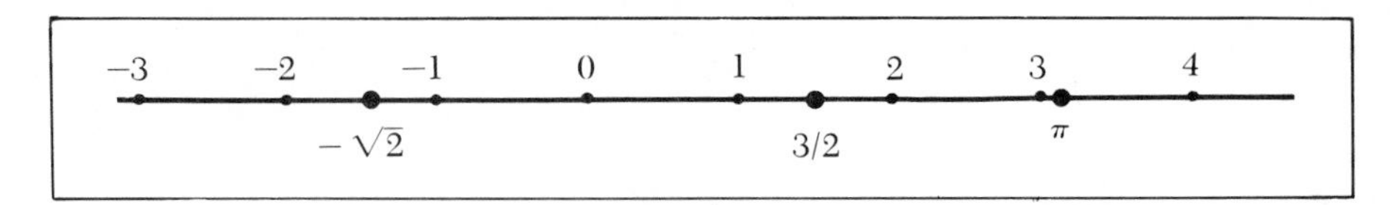

Figure 1–1 Each point on the straight line represents a number and each number is represented by a point. The numbers increase as we go from left to right.

The ordering of the real numbers can be represented as a straight line (Figure 1–1). This representation is usually arranged so that the numbers increase from left to right. The length of the line segment between 0 and 1 is defined as the *unit distance*. Any two successive integers are a unit distance apart. In fact, if a is any real number, then a is a unit distance from both $a - 1$ and $a + 1$; all the numbers lying between $a - 1$ and $a + 1$ are within distance 1 of a (Figure 1–2).

In the more general case, given any real number a and any positive real number p, we expect the "distance" from a to both $a - p$ and $a + p$ to be exactly p (Figure 1–3) and any numbers lying between $a - p$ and $a + p$ to be *within* distance p from a. The following definition provides us with a precise technique for computing the distance between any two real numbers.

Definition 4: *For any real number a, set*

$$|a| = \begin{cases} a, & \text{if } a \geq 0, \text{ and} \\ -a, & \text{if } a < 0. \end{cases} \tag{1}$$

We call $|a|$ the ***absolute value*** *of a. Given any two real numbers a and b, we define the* ***distance between them*** *to be $|a - b|$.*

NUMBERS DISTANCE 1 FROM *a*

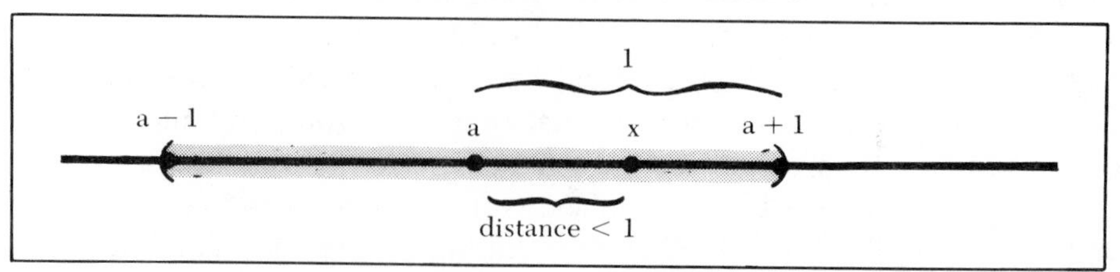

Figure 1–2 Any number x between $a - 1$ and $a + 1$ is within distance 1 of a.

NUMBERS WITHIN DISTANCE p OF a

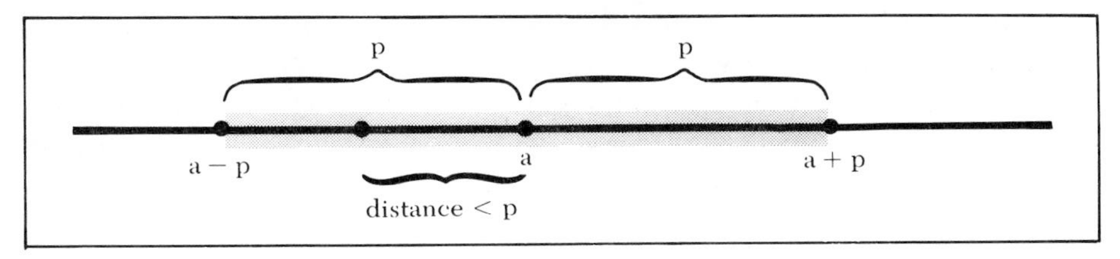

Figure 1–3 If x lies between $a - p$ and $a + p$, then x is within distance p of a.

The distance $|a - b|$ can be thought of as the length of the line segment joining a to b in the straight line representation of the real numbers.

The distance defined in Definition 4 has the properties we should expect a distance to have; if it did not, then it would fail to serve us in the manner intended, and the concept of distance between two real numbers would have to be redefined. We now list some properties that follow from the definition of distance. These should appear obvious from the context of the real number line.

(1) The distance between two real numbers a and b is 0 if and only if $a = b$. In symbols, $|a - b| = 0$ if and only if $a = b$.
(2) The distance from a to b is the same as the distance from b to a. In symbols, $|a - b| = |b - a|$.
(3) The distance from a to b plus the distance from b to c is at least as large as the distance from a to c. In symbols,

$$|a - b| + |b - c| \geqslant |a - c|.$$

(This is the so-called **Triangle Inequality.**)

The appropriateness of the absolute value as a measure of distance is illustrated by the next example.

Example 6: We concluded above that the set of numbers less than distance $p > 0$ from some number a is the set of numbers between $a - p$ and $a + p$. As a particular case, set $p = 1/2$ and $a = 5$. Then we expect the set of numbers less than distance 1/2 from 5 to be the numbers between $4\frac{1}{2}$ and $5\frac{1}{2}$, that is,

$$\{x \mid 4\tfrac{1}{2} < x < 5\tfrac{1}{2}\}.$$

If the absolute value is a suitable measure of distance, then this set should be the same as

$$\{x \mid |x - 5| < 1/2\},$$

which is simply the way of writing the set of numbers less than distance 1/2 from 5 using the absolute value notation. A suitable picture should convince us that $\{x \mid 4\frac{1}{2} < x < 5\frac{1}{2}\} = \{x \mid |x - 5| < \frac{1}{2}\}$ (Figure 1–4). We now provide a more formal argument that the sets in question are equal. To prove that the sets are equal we will show that the conditions which define them are equivalent; specifically, we will show that

$$4\tfrac{1}{2} < x < 5\tfrac{1}{2} \quad \text{if and only if} \quad |x - 5| < 1/2.$$

If x is some number between $4\frac{1}{2}$ and $5\frac{1}{2}$ ($4\frac{1}{2} < x < 5\frac{1}{2}$), then x can be expressed in the form $5 + p$ or $5 - p$, where p is a non-negative number less than 1/2. Therefore, $|5 - x|$ is equal to either

$$|5 - (5 + p)| = p < 1/2,$$

or

$$|5 - (5 - p)| = p < 1/2.$$

Consequently, the distance from x to 5 is less than 1/2. On the other hand, if x is some number within distance 1/2 of 5, then $|x - 5| < 1/2$. Now, $|x - 5| < 1/2$ means

$x - 5 < 1/2$ if x is greater than 5 and $-(x - 5) < 1/2$ if x is less than 5,

from which it follows that

$$x < 5 + 1/2 = 5\tfrac{1}{2} \quad \text{and} \quad x > 5 - 1/2 = 4\tfrac{1}{2}.$$

(A more complete discussion of computation involving inequalities and absolute values follows in the next section.) Therefore, if x is within distance 1/2 of 5, then $4\frac{1}{2} < x < 5\frac{1}{2}$.

We have just shown that if $4\frac{1}{2} < x < 5\frac{1}{2}$, then $|x - 5| < 1/2$; and that if $|x - 5| < 1/2$, then $4\frac{1}{2} < x < 5\frac{1}{2}$. Therefore, the two sets are equal.

A SET DEFINED IN TWO DIFFERENT WAYS

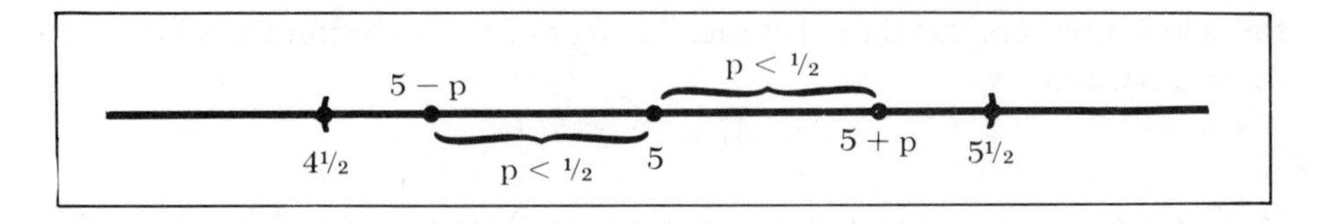

Figure 1–4 The numbers less than $5\frac{1}{2}$ but greater than $4\frac{1}{2}$, $\{x \mid 4\frac{1}{2} < x < 5\frac{1}{2}\}$, are precisely those numbers within distance $\frac{1}{2}$ of the number 5, $\{x \mid |x - 5| < \frac{1}{2}\}$.

The set of real numbers discussed in Example 6 is a special case of an important type of set of real numbers called an *open interval.*

Definition 5: *Let a and b be real numbers such that* $a < b$. *Then*

$$\{x \mid a < x < b\}$$

is called the **open interval** *with* **end points** *a and b. We denote this open interval by* (a, b).

The set

$$\{x \mid a \leq x \leq b\}$$

is called the **closed interval** *with end points a and b. (Note that the closed interval with end points a and b is* (a, b) *together with a and b.) We denote this closed interval by* $[a, b]$.

Observe that a square bracket indicates that an end point is to be included, whereas a round bracket indicates the exclusion of that end point.

With this distinction in mind, we naturally let $[a, b)$ denote the set

$$\{x \mid a \leq x < b\}, \tag{2}$$

and $(a, b]$ denote the set

$$\{x \mid a < x \leq b\}. \tag{3}$$

We call (2) and (3) *half-open intervals.*

Example 7: We see that $[-4, 3] = \{w \mid -4 \leq w \leq 3\}$, while $\{z \mid -1 \leq z \leq 1\} = [-1, 1]$.

We have already observed that if a is any real number and $p > 0$, then $\{x \mid |a - x| < p\}$ is the open interval $(a - p, a + p)$. In the next example we show that any open interval is a set of the type

$$\{x \mid |a - x| < p\}$$

for a suitable choice of a and p.

Example 8: Consider the open interval (a, b). The *midpoint* of this interval is $(a + b)/2$ (Figure 1–5). A justification of this assertion

THE MIDPOINT OF AN INTERVAL

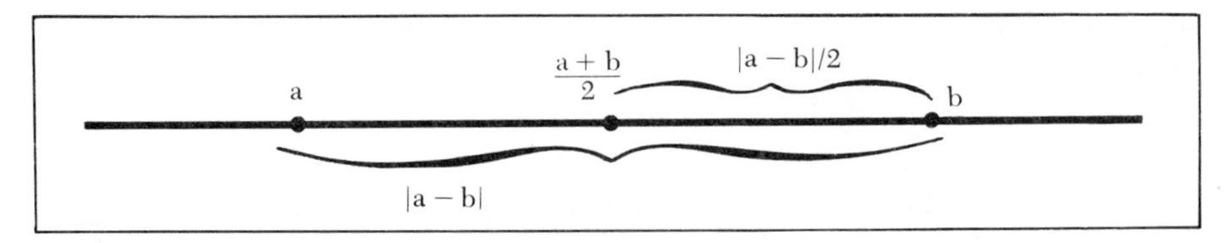

Figure 1–5 The point $(a + b)/2$ is distance $|a - b|/2$ from both a and b; $|a - b|$ is the distance from a to b, and hence $|a - b|/2$ is half the distance from a to b. Therefore, the midpoint of (a, b) is $(a + b)/2$.

proceeds as follows: If m is the midpoint of (a, b), then $m - a = b - m$; hence $2m = a + b$, or $m = (a + b)/2$. The distance from $(a + b)/2$ to each end point of (a, b) is simply $|a - b|/2$, half the length of (a, b). Therefore, (a, b) can also be thought of as

$$\{x \mid |\tfrac{1}{2}(a + b) - x| < \tfrac{1}{2}\,|a - b|\}.$$

Example 9: What is the set of numbers that approximate -6 with an accuracy of at least 0.01? This set consists of those numbers x that satisfy

$$|-6 - x| \leq 0.01,$$

and hence the set is $[-6.01, -5.99]$.

Example 10: Those real numbers within distance 1/4 of 5 are also within distance 1/2 of 5. The open interval $(4\frac{3}{4}, 5\frac{1}{4})$ is contained in the open interval $(4\frac{1}{2}, 5\frac{1}{2})$ (Figure 1–6). In other words, $(5 - 1/4, 5 + 1/4)$ is a subset of $(5 - 1/2, 5 + 1/2)$. More generally, if $0 < p < q$, then $(a - p, a + p)$ is a subset of $(a - q, a + q)$.

CONTAINMENT

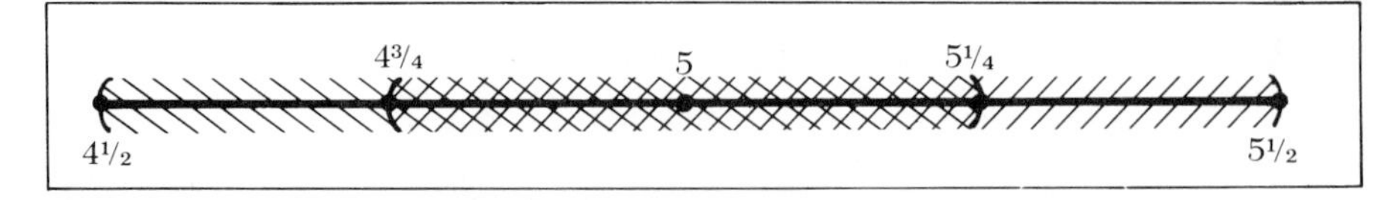

Figure 1–6 If x is within distance $\frac{1}{4}$ of 5, then x is also within distance $\frac{1}{2}$ of 5. Therefore, the open interval $(4\frac{3}{4}, 5\frac{1}{4})$ is contained in $(4\frac{1}{2}, 5\frac{1}{2})$.

Exercises

ROUTINE

1. Describe the sets of real numbers determined by each of the following conditions. Use the interval notation in your description when this is

possible. Represent each set pictorially on a straight line (see many of the figures used in this section).

a) $x > 3$ and $x < 7$
b) $-1 \leq x \leq 19$
c) $-4 < w \leq -1.9$
d) $|x - 3| < 1$
e) $|0.5 - x| < 1/2$
f) $|x + 1| < 1/4$
g) x approximates $\sqrt{2}$ with an accuracy of at least 0.1
h) $x < 1$ or $x > 3$
i) $x \geq \pi$
j) x is within distance 5 of 7.3
k) x is within distance -1 of 5

2. Find the union and intersection of the following sets.
   a) $\{x \mid x \text{ is a book}\}$; $\{y \mid y \text{ is red}\}$.
   b) $\{1, 2, 3\}$; $\{n \mid n \text{ is an odd integer}\}$.
   c) $\{n \mid n \text{ is an odd integer}\}$; $\{m \mid m \text{ is an even integer}\}$.
   d) $\{x \mid x \text{ is alive}\}$; $\{y \mid y \text{ is an American}\}$; $\{w \mid w \text{ owns a dog}\}$.

3. Each set below is equal to precisely one other set. Pair the equal sets.
   a) $\{x \mid 5 = x + 3\}$
   b) $\{x \mid 0 < x + 4 < 12\}$
   c) $\{x \mid |x| < 12 \text{ and } x > 4\}$
   d) $\left\{x \,\middle|\, \dfrac{x+2}{2} = 3\right\}$
   e) $(4, 9] \cup (7, 12)$
   f) $\{x \mid x^2 = 4\}$
   g) $\{x \mid |x - 5| < 12\}$
   h) $[-12, 24) \cap (-10, 36)$
   i) $\{x \mid |x - 5| = |3 - x|\}$
   j) $\{x \mid 17 > |x - 7|\}$
   k) $\{x \mid -3 < x < -5\}$
   l) $\{x \mid -7 < x < 17\}$
   m) $[-6, 12) \cap [16\frac{1}{2}, 24]$
   n) $\{x \mid (x - 2)^2 = 0\}$
   o) $\{x \mid |x| = 2\}$
   p) $(-4, 8)$

CHALLENGING

4. Prove that the two sets in each of the following pairs of sets are equal. Do this by showing that a number satisfies the condition to be in one of the sets if and only if it satisfies the condition to be in the other.
   a) $\{x \mid x^2 = 1\}$; $\{-1, 1\}$
   b) $\{x \mid 2 \leq x \leq 4\}$; $\{x \mid |3 - x| \leq 1\}$
   c) $\{x \mid x \text{ is brown}\} \cap \{y \mid y \text{ is a book}\}$; $\{w \mid w \text{ is a brown book}\}$
   d) $(0, 1] \cup [1, 2)$; $(0, 2)$

5. Give an argument to demonstrate the truth of each of the following statements:
   a) Any set is a subset of itself.
   b) If $S \subset T$ and $T \subset W$, then $S \subset W$.
   c) $S \cap T$ is a subset of both S and T.
   d) S and T are both subsets of $S \cup T$.

6. Prove that $\sqrt{2}$ is irrational. [Hint: Suppose $\sqrt{2}$ is rational. Then there are positive integers m and n such that $(m/n)^2 = m^2/n^2 = 2$. We may assume that m and n are not both divisible by 2. Show that if m^2 is divisible by 2, then m is also divisible by 2. Then show that m and n are divisible by 2, which contradicts our assumption.]

7. Prove that $\sqrt{3}$ is irrational.

THEORETICAL

8. For each of the following interpretations of A, B, C, and D, find $A \cup B$, $A \cap C$, $B \cup D$, and $C \cap D$. Is $A \subset B$? Is $D \subset A$? Is $C \subset B$?
   a) A = dorms, B = coed dorms, C = brick buildings, D = buildings containing beds
   b) A = teenagers, B = mothers, C = daughters, D = females
   c) A = conjunctions, B = {and}, C = words that begin with "a," D = words that begin with a vowel
   d) A = California cities, B = {Chicago}, C = {Los Angeles}, D = {San Francisco}
   e) A = sons, B = sisters, C = brothers, D = people under age 20

1.2 COMPUTATIONS WITH ABSOLUTE VALUES AND INEQUALITIES

Although equations such as

$$x - 6 = 9$$

are encountered early in most high school algebra courses, equations such as

$$|x - 6| = 9$$

may not have been encountered before reading this text. The same observation applies to inequalities such as

$$x - 6 < 9 \quad \text{or} \quad x - 6 \leq 9.$$

For this reason, we now present a section which reviews the basic methods of solving equations involving absolute values, as well as solving various types of inequalities.

Solving Equations Involving an Absolute Value

In solving the equation

$$|x - 6| = 9 \tag{4}$$

we must remember that the definition of $|x - 6|$ depends on whether $x - 6$ is non-negative or negative. If $x - 6$ is non-negative, then $|x - 6|$ is equal to $x - 6$ and equation (4) takes the form

$$x - 6 = 9,$$

from which we obtain $x = 15$. On the other hand, if $x - 6$ is negative, then $|x - 6| = -(x - 6) = 6 - x$, and equation (4) takes the form $-(x - 6) = 9$, or

$$6 - x = 9,$$

from which we obtain $x = -3$. Both 15 and -3 are solutions to equation (4). One solution is obtained by assuming that $x - 6$ is non-negative, while the other solution is obtained by assuming that $x - 6$ is negative.

In general, to solve an equation of the form

$$|x - a| = b, \tag{5}$$

where a and b are constants, we must solve both $x - a = b$ and $-(x - a) = b$. The solutions to these equations are $x = b + a$ and $x = b - a$. If b is non-negative, both of these numbers are solutions to equation (5) and, in fact, are *all* of the solutions to equation (5). If b is negative, then equation (5) has no solutions because an absolute value can never be negative.

Rules Governing the Arithmetic of Inequalities

The rules given below are stated in terms of strict inequality ($<$). They apply equally well when $\leq$ (less than or equal to) is substituted for $<$ (less than).

A. If $a < b$ and c is any number, then $a + c < b + c$.

For example, $3 < 5$ implies $3 + 10 = 13 < 5 + 10 = 15$ and $3 + (-2) = 1 < 5 + (-2) = 3$.

B. If $a < b$ and c is any *positive* number, then $ac < bc$.

For example, $7 < 8$ implies $7 \cdot 2 = 14 < 8 \cdot 2 = 16$ and $-3 < 4$ implies $(-3)7 = -21 < 4 \cdot 7 = 28$.

C. If $a < b$ and c is any negative number, then $bc < ac$. (Note that multiplying both sides of an inequality by a negative number reverses the inequality.)

For example, $7 < 8$ implies $8(-2) = -16 < 7(-2) = -14$ and $-3 < 4$ implies $4(-7) = -28 < (-3)(-7) = 21$.

As a special case of rule C, we see that $a < b$ implies $b(-1) = -b < a(-1) = -a$.

D. If a and b are positive numbers with $a < b$, then $1/b < 1/a$

(This rule simply says that the larger a positive number is, the smaller is its reciprocal.)

For example, since $18 < 54$, we can conclude $1/54 < 1/18$.

We make no attempt to prove the rules governing inequalities; if properly understood, they should appear virtually as common sense.

Solutions of Inequalities Involving One Variable

The linear equation $x - 6 = 9$ has precisely one solution, while the equation $|x - 6| = 9$ has two solutions. The inequality

$$x - 6 < 9 \tag{6}$$

has infinitely many solutions. By adding 6 to both sides of inequality (6) and using rule A, we see that a number x satisfies inequality (6) provided $x < 15$. Any number less than 15 satisfies inequality (6)—for example, 14, 13.6, -90, and $-6/5$ are all solutions—and there are infinitely many numbers less than 15. We see then that the set of numbers which satisfy an inequality may be very large.

Example 11: We use the rules governing inequalities to solve

$$1/x - 3 \leq -6. \tag{7}$$

We first add 3 to both sides of inequality (7) to obtain

$$1/x \leq -3. \tag{8}$$

If x is to be a positive solution, then we can multiply both sides of inequality (8) by x according to rule B and obtain

$$1 \leq -3x. \tag{9}$$

Then, multiplying both sides of inequality (9) by $-1/3$ in accordance with rule C, we obtain

$$x \leq -1/3.$$

We therefore conclude that any *positive* number x which is a solution to inequality (7) must also be less than or equal to $-1/3$. There is no such number. Thus, inequality (7) has no positive solutions.

We now determine whether inequality (7) has any negative solutions. Assume x is negative and multiply inequality (8) on both sides by x. In accordance with rule C, we thus obtain

$$-3x \leq 1. \tag{10}$$

Now, multiplying both sides of inequality (10) by $-1/3$, we find

$$-1/3 \leq x. \tag{11}$$

Therefore, any negative solution x for inequality (7) must be greater than or equal to $-1/3$. Since to say that x is negative is the same as to say $x < 0$, a solution to inequality (7) is any number less than 0 but greater than or equal to $-1/3$. In other words, the solutions to (7) are found in the interval $[-1/3, 0)$.

Simultaneous Inequalities

If we think of an equality or inequality as a condition that a number must satisfy, then the desire to impose several conditions simultaneously on some number may lead us to require that a number be a solution of not merely one but of several inequalities at the same time.

Example 12: A manufacturer of toothbrushes estimates that a quality brush cannot be produced for less than \$0.15, but market conditions also indicate that a brush which costs more than \$0.40 to produce would not attract enough of the retail market to justify its production. If x represents the cost in cents of each brush produced, then in order to have a quality competitive product, the manufacturer will demand that x satisfy the inequalities

$$15 < x$$

and

$$x < 40$$

simultaneously.

To find those numbers which satisfy all of the members of a set of inequalities, we solve each inequality separately and then compare the solutions of the individual inequalities to find those numbers which satisfy all of them together. In set theoretic terms, we find the set of solutions to the simultaneous inequalities by taking the intersection of the solution sets for each of the individual inequalities.

Example 13: We will find all numbers which simultaneously satisfy

$$x - 5 < -1 \tag{12}$$

and

$$3 - x \leq 6. \tag{13}$$

In order to satisfy inequality (12), x must be less than 4; thus the solution set of (12) is $\{x \mid x < 4\}$. In order to satisfy inequality (13), x must be greater than or equal to -3; thus the solution set for (13) is $\{x \mid -3 \leq x\}$. In order to satisfy inequalities (12) and (13) simultaneously, x must be both less than 4 and greater than or equal to -3; thus the solution set to (12) and (13) together is $\{x \mid -3 \leq x < 4\}$ (Figure 1–7).

SOLUTION SET FOR SIMULTANEOUS INEQUALITIES

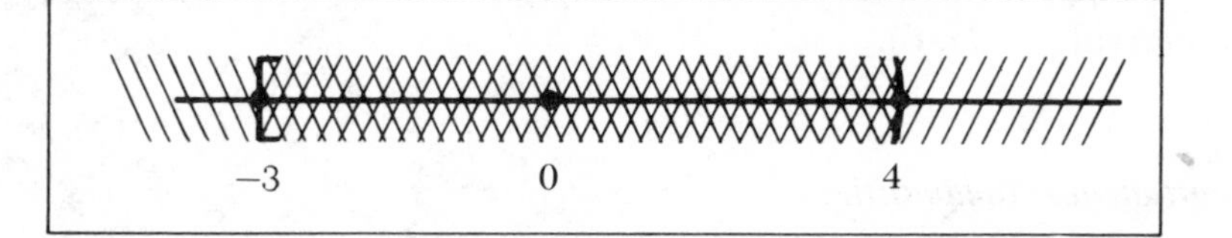

Figure 1–7 The solution set for inequality (12) is $\{x \mid x < 4\}$; the solution set for inequality (13) is $\{x \mid x \geq -3\}$. The intersection of these two sets is $[-3, 4)$; this is the solution set for (12) and (13) simultaneously.

It is possible that a set of inequalities may not be satisfied by any real number; for example, the inequalities

$$x < -1 \quad \text{and} \quad 2 < x$$

have no simultaneous solutions, since a number cannot be less than -1 and greater than 2 at the same time.

Inequalities Involving Absolute Values

We already noted that if $p > 0$, then the set of all numbers x which satisfy

$$|x - a| < p \tag{14}$$

is the same as the set of numbers x which lie within distance p of a, which, in turn, is simply the open interval $(a - p, a + p)$. We now indicate how we actually solved inequality (14) in order to arrive at our conclusion.

Because the definition of the absolute value involves two cases, we split the solution of inequality (14) into two cases. First, if $a \leq x$, then $x - a$ is non-negative and $|x - a| = x - a$. Thus, for $a \leq x$, (14) becomes the inequality

$$x - a < p. \tag{15}$$

In order to satisfy inequality (15), x must be less than $a + p$. But to obtain (15), we also assumed $a \leq x$. Thus, $\{x \mid a \leq x < a + p\}$ forms part of the solution set for inequality (14).

If, on the other hand, we take $x < a$, then $x - a$ is negative and $|x - a| = -(x - a) = a - x$. This, in turn, leads to the inequality

$$a - x < p, \tag{16}$$

and the solution set $\{x \mid a - p < x < a\}$. The solutions for inequality (14) then comprise all the numbers from $a - p$ to a, together with all the numbers, including a, from a to $a + p$. The total solution set for (14) is then $(a - p, a + p)$; this is shown in Figure 1–8.

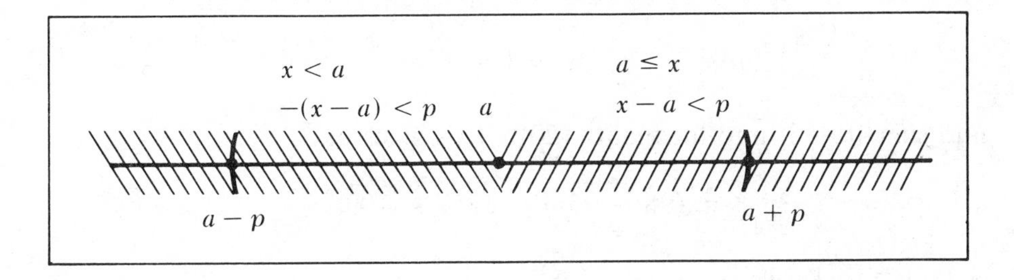

Figure 1–8 To find the solution set for $|x - a| < p$ we "add" the solution sets for the two cases $x < a$ and $a \leq x$.

Example 14: We will solve

$$|4x + 7| > 5. \tag{17}$$

If we set $y = 4x$, then we can write inequality (17) as

$$|y - (-7)| > 5. \tag{18}$$

The solution set for inequality (18) consists of all numbers y greater than distance 5 from -7. Now we note that the numbers with distance less than or equal to 5 from -7 make up the closed interval $[-12, -2]$. Hence, in order to satisfy inequality (18), y must be outside $[-12, -2]$; that is, y must be less than -12 or greater than -2. Since $y = 4x$, the solution set for inequality (17) is $\{x \mid 4x < -12, \text{ or } 4x > -2\} = \{x \mid x < -3, \text{ or } x > -1/2\}$.

Inequalities Involving Two Unknowns

Often some relationship between two quantities is appropriately expressed by means of some inequality.

Example 15: A confectioner decides to produce a mixture of two types of nuts; he also determines that the mixture must be produced for less than \$1.09 per pound. If the two types of nuts to be mixed cost \$0.95 and \$1.25 per pound, respectively, and are used in quantities of x and y pounds to form a one pound mixture, then x and y must satisfy

$$x + y = 1 \tag{19}$$

and

$$95x + 125y < 109, \qquad x \geq 0, \qquad y \geq 0. \tag{20}$$

To obtain a solution, we first rewrite equation (19) in the form

$$x = 1 - y$$

and substitute it into inequality (20). This gives us

$$(95 - 95y) + 125y < 109.$$

Adding like terms, we obtain

$$30y < 14$$

or

$$y < \tfrac{7}{15}.$$

Adding to this the given restraint that $y \geq 0$, we obtain as the solution set in y the set $\{y \mid 0 \leq y < 7/15\}$. The accompanying solution set in x is $\{x \mid 8/15 \leq x \leq 1, \text{ and } x = 1 - y\}$. This can be written more compactly as

$$\{(x, y) \mid 0 \leq y < 7/15, \quad \text{and} \quad x = 1 - y\}.$$

Example 16: A condition that the sum of two numbers x and y be less than 1 can be written

$$x + y < 1. \tag{21}$$

A solution for inequality (21) is an ordered pair of numbers (x, y) such that when x and y are substituted in (21) the given inequality is satisfied. For example, (1/3, 1/4) is a solution for (21), while (2, 3) is not a solution. The *solution set* of (21) consists of all pairs (x, y) which are solutions for (21). In a later section we will introduce a geometric device for picturing the solution sets for inequalities involving two unknowns.

This geometric device will also prove a convenient tool in the simultaneous solution of several inequalities involving more than one unknown.

Exercises

ROUTINE

1. Assume $a < b$. Which of the following are always true? Which are only sometimes true; in other words, which may be true or false depending on the values of a and b? Which are never true?

a) $5a < 5b$
b) $-3a > -3b$
c) $a - 5 < b - 5$
d) $3a - 5 \leq 3b - 5$
e) $1/a > 1/b$
f) $a^2 < b^2$
g) $|a| < |b|$
h) $a - b > 0$
i) $3a < 5b$
j) $-3b < -5a$
k) $\frac{1}{2}a - 3 < \frac{1}{2}b - 2$
l) $|a - 5| < |b - 5|$
m) $a^3 < b^3$

2. Solve each of the following.

a) $x - 6 < 3$
b) $|2x - 6| = 3$
c) $2x + 6 < 3$
d) $2x + 6 \geq 3$
e) $3x - 1 < 5$
f) $-3x - 1 < 5$
g) $|3x - 17| < 5$
h) $4x - 16 > -18$
i) $|x - 3| \leq 10$
j) $1/x - 16 < -3$
k) $3/x + 5 < 1$
l) $|1/x - 1| < 1$
m) $(4x - 1)(2x - 1) > 0$ (Hint: For the product of two numbers to be positive, either both numbers must be positive, or both numbers must be negative.)
n) $10 < |3x + 9|$
o) $1/x < x$

3. Find the simultaneous solutions to the equations and inequalities in each of the following.

a) $3x \geq 1$
$4x - 1 < 5$
b) $|x - 8| = 9$
$4x < 100$
c) $x > 0$
$1/x < 1$
$4x - 2 \leq 3$
d) $|2x - 3| < 5$
$x \leq |3x - 1|$
e) $x^2 > 1$
$|x - 9| < 54$
$x - 6 > -3$

CHALLENGING

4. Express each of the following as an appropriate inequality, or set of inequalities.

a) A mechanic expects to spend x hours on one job and y hours on another, but he does not want to spend more than 8 hours total time on both jobs.
b) A contractor will hire a bulldozer for x hours at \$35 per hour and a power shovel for y hours at \$50 per hour, but he does not expect the total cost of the project to exceed \$4500.
c) The sum of the squares of two numbers is less than 1/5 of their sum.

1.3 FUNCTIONS. THE COORDINATE PLANE

Henceforth, we will let R denote the set of real numbers, and unless we specify otherwise, any set under consideration will be a set of real numbers.

We often try to determine one quantity, given another quantity. For example, given the distance to be travelled from one city to another, we might estimate the number of gallons of gas it would take to drive that distance, or the time in hours required to make the trip. Given the college board scores of an enrolling student, the registrar of a college might try to predict the overall grade point average the student will achieve in college.

To make predictions or estimates such as these, one generally employs a formula or rule that establishes a relationship between the given quantity and the quantity to be determined. For example, if a person knows that his car averages 16 miles per gallon, then he will figure the number g of gallons of gasoline needed to make a trip of x miles, using the equation

$$g = x/16$$

Again we demonstrate how one quantity can be computed once another quantity is given.

Example 17: A piece goods worker receives exactly \$0.05 for each item he produces. If d is the pay earned by the worker in one day, and x is the number of items he produces on that day, then x and d are related by the equation $d = 0.05x$. Once the number of pieces the worker produced is known, we can compute precisely the wages earned.

Sometimes, too, we are given pairs of numbers that measure related quantities. In such an instance there may be a rule relating the two quantities, and our goal is to find that rule.

Example 18: A shopper knows the wholesale prices of certain items, and he observes the retail prices of these items in a particular store. The shopper then constructs a table (Table 1.1).

TABLE 1.1

Wholesale Price	*Retail Price*
0.90	0.99
1.50	1.65
0.50	0.55
2.00	2.20

On the basis of these observations, the shopper attempts to find a relationship between the wholesale and retail prices of any item. Letting x and y be the wholesale and retail prices, respectively, of any item, the shopper sets up, on the basis of Table 1.1, the equation

$$y = x + 0.1x.$$

Still another example in which two quantities have a precise relationship to one another is the following.

Example 19: The volume of a cube with sides of length x is given by

$$V = x^3.$$

The relationship between V and x is clearly more complicated than the relationship between d and x in Example 17, although the former is still relatively simple.

In each of the preceding three examples, it is possible, given one number, to calculate precisely another real number that depends on the given one in a clearly defined way. Given x in Example 17, we can immediately compute d. Similarly, given x in Example 19, we can compute V precisely; once x is specified in Example 18, y can be determined using the equation $y = x + 0.1x$.

This situation, in which one quantity can be found once some other quantity is known, is of central importance to much of mathematics and to the practical applications of mathematics.

Definition 6: *A rule, phrase, relationship, or other device that associates each element of one set S with one and only one element of a set T is called a* ***function from S into T.*** *(Remember that for our purposes, we are assuming S and T to be sets of real numbers. This definition, however, is valid for any sets S and T.)*

A function f from S into T is usually described by specifying for each s in S the element $f(s)$ in T that f associates with s. Thus, in Example 17 we can write

$$d(x) = 0.05x$$

to indicate that d is a function of x. Given the numerical value of x, we can compute $d(x)$, the number $0.05x$. Likewise, in Example 19 we can write $V(x) = x^3$ to indicate that V is a function of x and to indicate the precise manner in which V is calculated once x is given.

> **Definition 7:** *If f is a function from S into T, then for each element s of S, we call $f(s)$ the* ***function value of f at s;*** *we call s the* ***variable****, or* ***argument****, of the function. We call S the* ***domain*** *of f, and T the* ***range*** *of f.*

This particular definition of the range of a function is not the only one. The range may be defined as the subset of T that consists of all the function values of f. We will, of course, be consistent in applying any definition we introduce, but, unfortunately, there is no guarantee that usage of a particular term will be uniform among different authors.

If f is a function from S into T, then each element of S has a function value $f(s)$. It is not, however, necessarily true that every element of T is a function value, or that an element of T is a function value for at most one element of S (Figure 1–9).

RANGE OF A FUNCTION

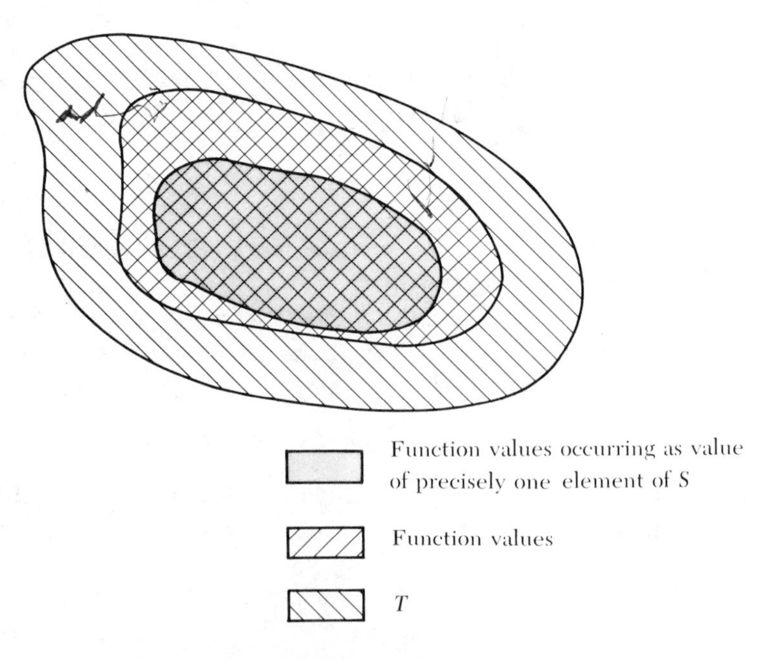

Figure 1–9 Not all elements of T need be function values; even those elements of T which are function values may be function values for more than one element of S.

Example 20: Let f be the function from R into R defined by

$$f(x) = x^2;$$

remember that R represents the set of real numbers. Here the domain and range of f are both R. Note that no negative real number occurs as a function value for f, because x^2 is always non-negative for any real number x (Figure 1–10). In this case two different real numbers can have the same function value; for example, $f(1) = 1^2 = 1 = (-1)^2 = f(-1)$.

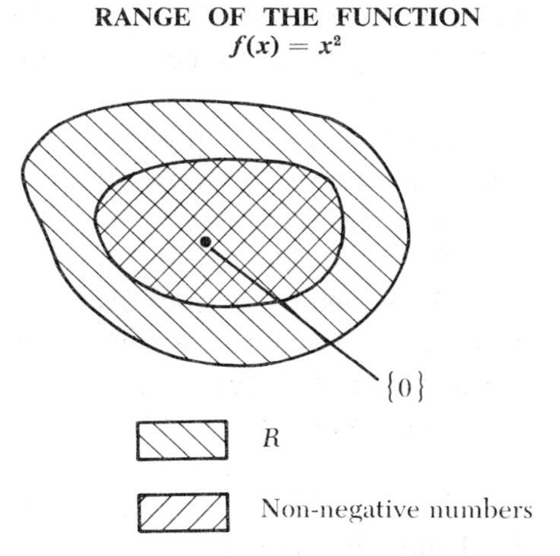

Figure 1–10 Only non-negative numbers occur as function values for f; 0 is the only number which is the function value of only one real number.

Example 21: Let g be a function from R into R defined by $g(x) = 2x$. Each real number occurs as a function value for g, because if y is any real number, then $g(y/2) = 2(y/2) = y$. Hence y is the function value for $y/2$ (Figure 1–11). Moreover, two distinct real numbers will have two distinct function values; that is, if $x \neq x'$, then $g(x) = 2x \neq 2x' = g(x')$.

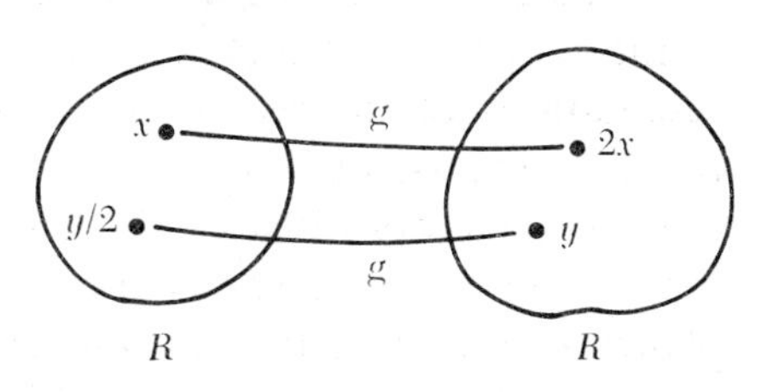

Figure 1–11 Any real number y occurs as a function value of g once and only once; specifically, $y = g(x)$ if and only if $x = y/2$.

We have specified that both the domain and the range of any function under consideration will be sets of real numbers. This restriction was imposed because functions of this sort are solved using the techniques of the calculus. We will also assume the domain of any function to be the largest set of real numbers for which the definition of the function is valid. Thus, the domain of the function will not generally be stated explicitly but will be determined by the logical or physical limitations of the particular problem.

Example 22: In Example 17, the function variable x is the number of items the worker produces. This number must be a non-negative integer; hence the nature of the problem leads us to consider the set of non-negative integers as the domain of the function $d(x)$.

Example 23: In Example 19, the function variable is the length of any side of the cube. Since a length cannot be negative, the domain of the function $V(x)$ will be the set of non-negative real numbers.

Example 24: Suppose $f(x) = 1/x$. The quotient $1/x$ is defined for every real number x except 0; therefore, the domain of f will be the set of non-zero real numbers.

Occasionally we may override the convention introduced above and restrict the domain of a function to a set smaller than the one our convention might dictate. In such cases we will always indicate exactly what the domain is to be.

Example 25: Let the function f from the open interval (0, 1) into R be defined by $f(x) = x^2$. Here we specifically restrict the domain of f to (0, 1) even though $f(x)$ has a value for any real number x.

It is sometimes useful to represent a function graphically; the *coordinate plane* is ordinarily used to graph functions of the type we are studying.

In order to set up a system of coordinates in the Euclidean plane, we construct two perpendicular lines, one called the *x-axis* for *x-values*, and one called the *y-axis* for *y-values*, with positive directions as shown in Figure 1–12. Note that we are essentially placing two straight line representations of the real numbers at right angles to one another. Each point on the x-axis has *coordinates* of the form $(x, 0)$, and each point on the y-axis has coordinates of the form $(0, y)$. The point of intersection of the x- and y-axes is called the *origin* and has coordinates (0, 0). In the plane, a point P that does not lie on either axis is assigned coordinates according to the method depicted in Figure 1–12.

We assume some familiarity with the coordinate plane. This description of its construction is presented for the sake of review. In the following sections we will see how the coordinate plane can be used to represent pictorially functions, inequalities, and certain other concepts pertinent to the real numbers.

THE COORDINATE PLANE

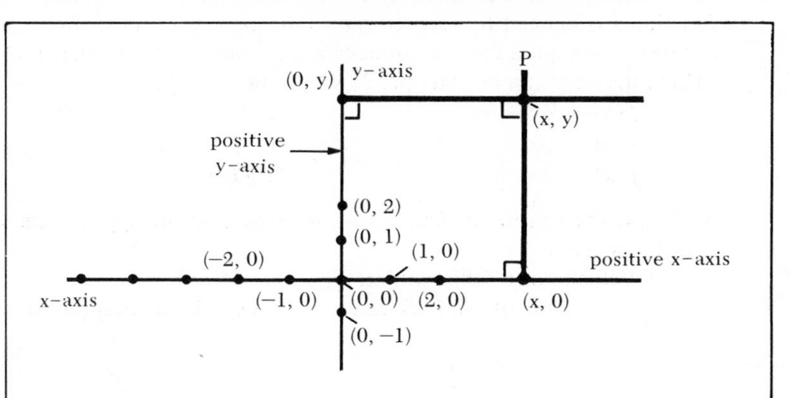

Figure 1–12 The coordinate plane is useful for graphing equations and functions. The line parallel to the x-axis through the point P intersects the y-axis at $(0, y)$. The line parallel to the y-axis through P intersects the x-axis at $(x, 0)$. The coordinates of P are (x, y).

Exercises

ROUTINE

1. Determine the domain of each of the following functions. Find the function value at 2 if 2 is in the domain of the function, and at 0 if 0 is in the domain of the function.
   a) $f(x) = x$
   b) $g(w) = w^2$ (Note that x need not be used to represent the function variable, nor does f have to be used to represent a function.)
   c) $h(t) = 4t - 1$
   d) $f(x) = x^3$
   e) $m(t) = \sqrt{t - 1}$
   f) $d(t) = 4t^3 - 1$
   g) $f(w) = -w - 8$
   h) $r(x) = 1/(x - 1)$
   i) $g(y) = 1/y^2$
   j) $k(z) = 1/(z^2 + 1)$

2. (a) Which functions in Exercise 1 have the property that no two distinct values of the function variable yield the same function value?
   (b) Which functions in Exercise 1 have every real number as a function value?

3. Locate each of the following points in the x-y coordinate plane.
   a) $(1, 1)$
   b) $(1, -5)$
   c) $(-1, -5)$
   d) $(-1, 5)$
   e) $(-1/2, 1/4)$
   f) $(\sqrt{2}, 0)$
   g) $(2 + \sqrt{2}, 2 - \sqrt{2})$
   h) $(0.3, -0.7)$

4. In Example 21 we saw that for any real number y, $g(y/2) = y$. A function f is defined by each of the following expressions. For any real number y, compute the real number z (in terms of y) such that $f(z) = y$. (Thus, the equation in Example 21 would be $z = y/2$.)

a) $f(x) = 2x - 1$
b) $f(x) = 4x$
c) $f(x) = x + 8$
d) $f(x) = -6x + 10$
e) $f(x) = \sqrt{x}$
f) $f(x) = x^3$

5. Graph the following data, using in each case an appropriate set of coordinate axes.

a) Velocity of a falling object

Time of fall in Seconds	Velocity in Feet per Second
0	0
1	32
2	64
3	96

b) Approximate population of the world

Date	Population in Millions
1650	550
1750	725
1800	900
1850	1175
1900	1600
1950	2490
1965	3237

c) Corporate profit

Date	Profit in Thousands of Dollars
1960	3.0
1961	3.5
1962	−0.5
1963	2.5
1964	4.0
1965	5.6
1966	5.9
1967	7.2
1968	7.5
1969	6.5
1970	5.4

CHALLENGING

6. Graph the points $(-3/4, -3/2)$, $(-1/3, -2/3)$, $(1/2, 1)$, $(1/2)$, and $(\sqrt{2}, 2\sqrt{2})$. For each point, what is y in terms of x? Graph additional points that fit the pattern.

1.4 GRAPHS

One of the fundamental concepts in plane geometry is that of a straight line. This concept plays a central role in our development of the calculus. Let us now investigate straight lines in the coordinate plane.

Consider a line parallel to the x-axis (Figure 1–13). Each point on

POINTS WITH y-COORDINATE a

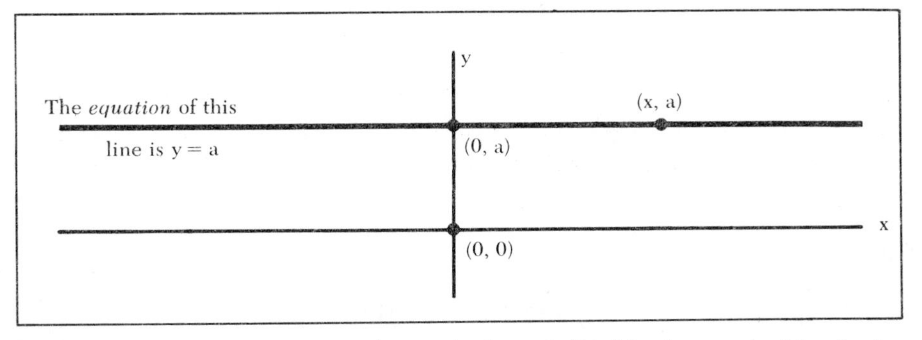

Figure 1–13 The line parallel to the x-axis through $(0, a)$ is characterized by the fact that each point on it has y-coordinate a and all points with y-coordinate a lie on the line.

this line will have coordinates of the form (x, a), where $(0, a)$ is the point where the line intersects the y-axis. Moreover, each point in the plane with coordinates of the form (x, a) will lie on this line. This follows as an immediate consequence of the way in which coordinates are determined for any point in the plane. Thus, the line that is parallel to the x-axis and that intersects the y-axis at the point $(0, a)$ consists of all points (x, y) such that $y = a$. We call $y = a$ the *equation* of this straight line.

Similarly, if a line parallel to the y-axis intersects the x-axis at the point $(a, 0)$ as in Figure 1–14, then we can readily show that this line consists of all points (x, y) such that $x = a$. The equation of this line is $x = a$.

Suppose now that a straight line L passes through two distinct points (x_0, y_0) and (x_1, y_1), but is not parallel to the y-axis (Figure 1–15). We now find an equation for L.

Suppose that (x, y) is any point on L, with $(x, y) \neq (x_1, y_1)$. Because L is a straight line, the two right triangles pictured in Figure 1–15 are similar. This means that the ratios of the lengths of the corresponding sides are equal. In particular, we have

$$\frac{y - y_1}{x - x_1} = \frac{y_1 - y_0}{x_1 - x_0} \tag{22}$$

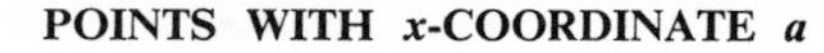

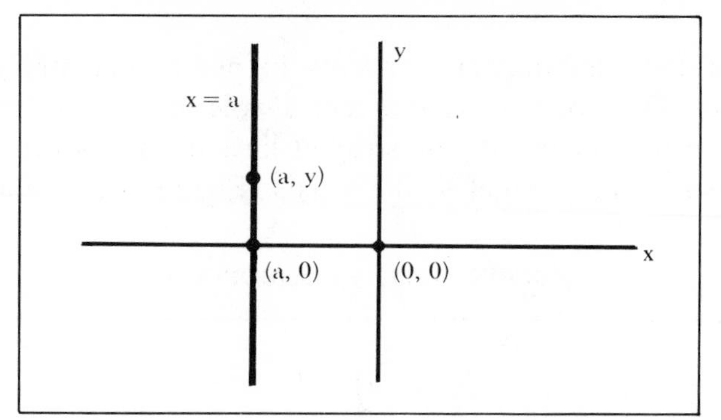

Figure 1–14 The line parallel to the y-axis through $(a, 0)$ is characterized by the fact that each point on it has x-coordinate a and all points with x-coordinate a lie on the line.

or

$$y - y_1 = \left(\frac{y_1 - y_0}{x - x_0}\right)(x - x_1). \tag{23}$$

We observe that if (x, y) is any point on the line L, then x and y will satisfy equation (23). We can also show that if x and y satisfy (23), then (x, y) is a point of L. Therefore (23) is the equation of L, the line determined by (x_0, y_0) and (x_1, y_1).

Note that equation (22) is not satisfied by (x_1, y_1) since with $x = x_1$ and $y = y_1$, the left side of (22) is the undefined expression 0/0. To be sure

EQUATION OF THE LINE THROUGH TWO DISTINCT POINTS

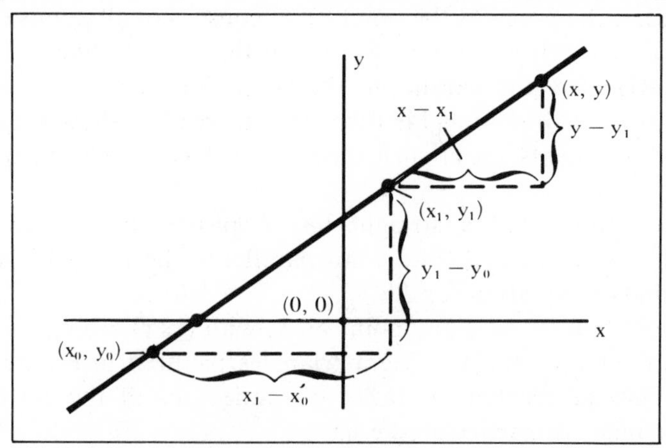

Figure 1–15 If the line L passes through the points (x_0, y_0) and (x_1, y_1), with $x_0 \neq x_1$ then the equation for L is $y - y_1 = \dfrac{y_1 - y_0}{x_1 - x_0}(x - x_1)$. The line L consists of precisely those points (x, y) which satisfy this equation.

of including all of the points of L, and (x_1, y_1) in particular, we must use form (23) as the equation of the line.

Example 26: Determine the equation of the line passing through the points $(-1, 4)$ and $(2, 3)$. Letting $(-1, 4)$ and $(2, 3)$ represent (x_0, y_0) and (x_1, y_1), we substitute these values in equation (23) and obtain

$$y - 3 = \left(\frac{3 - 4}{2 - (-1)}\right)(x - 2) = (-1/3)(x - 2)$$

as the equation of the line in question (see Figure 1–16). A point (x, y) of the plane will lie on this line if and only if the x and y values satisfy the equation of the line, $y - 3 = (-1/3)(x - 2)$.

GRAPH OF THE LINE THROUGH (−1, 4) AND (2, 3)

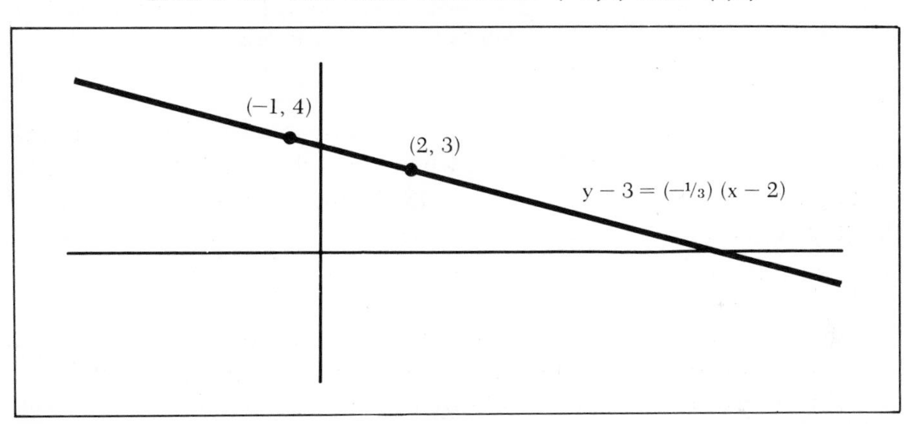

Figure 1–16 We use equation (23) to determine that the equation for this line is $y - 3 = (-\frac{1}{3})(x - 2)$.

We call equation (23) the *two-point* form of the equation of a straight line.

Let us now examine the quotient $(y_1 - y_0)/(x_1 - x_0)$ on the right side of equation (22). This quotient is, of course, a number; we shall call it the *slope* of L. The slope was computed using the points (x_0, y_0) and (x_1, y_1), but, considered geometrically, the slope is the tangent of the angle α in Figure 1–17. Since the two right triangles are similar, angle α = angle β, and therefore the tangents of these two angles are equal. Hence, if we choose *any* two distinct points (a, b) and (c, d) of L, the quantity $(d - b)/(c - a)$ will be the same regardless of the choice of points. In other words, any two distinct points of L can be used to compute the slope of L. We can therefore associate the slope of L with L itself rather than with any particular pair of points on L.

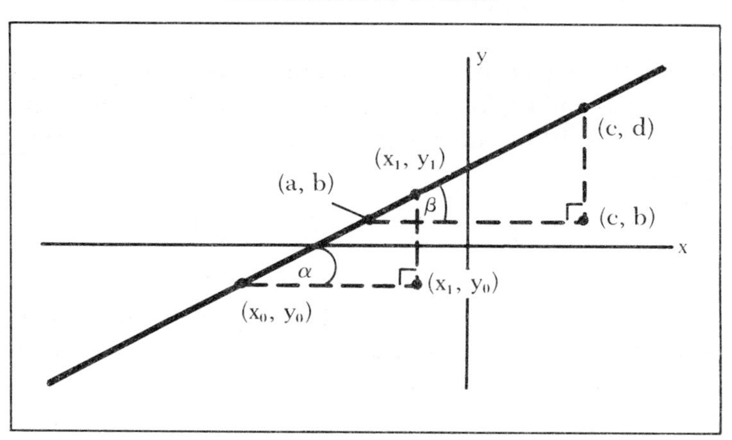

Figure 1–17 Any two right triangles whose hypotenuses lie on L will be similar. Therefore, the ratio $(d - b)/(c - a)$ will be the same regardless of which points (a, b) and (c, d) of L are used to compute it.

If L is a line parallel to the x-axis, its slope is zero. Such a line will have an equation of the form $y = k$, where k is a constant. Consequently, for any two points of L—for instance, $(3, k)$ and $(7, k)$—we can compute the slope of L:

$$(k - k)/(3 - 7) = 0/(-4) = 0.$$

If L is a line that intersects the x-axis at the point $(a, 0)$ and the y-axis at $(0, b)$, then the slope of L is $b/(-a) = -b/a$. This number is also the tangent of the angle made by the line and the x-axis, the angle being measured counterclockwise from the x-axis to the line (see Figure 1–18).

It follows from the nature of the slope of a line that two lines will have the same slope or will both be parallel to the y-axis (in which case their slopes are not defined) if the two lines are parallel. Conversely, two lines will be parallel if they have the same slope or are both parallel to the y-axis.

Any given real number is the slope of some line. Lines of positive slope form an acute angle with the positive x-axis; the larger the slope, the larger the angle. Lines with 0 slope are parallel to the x-axis, and lines with negative slope form an obtuse angle with the x-axis (Figures 1–19 and 1–20).

In specifying the slope of a line, we do not determine the line completely, since any line parallel to the given one will have the same slope. However, by specifying both the slope of a line L and a point through which the line passes, we determine L uniquely; L will be the only line that has the given slope and passes through the given point.

LINE WITH SLOPE $-b/a$

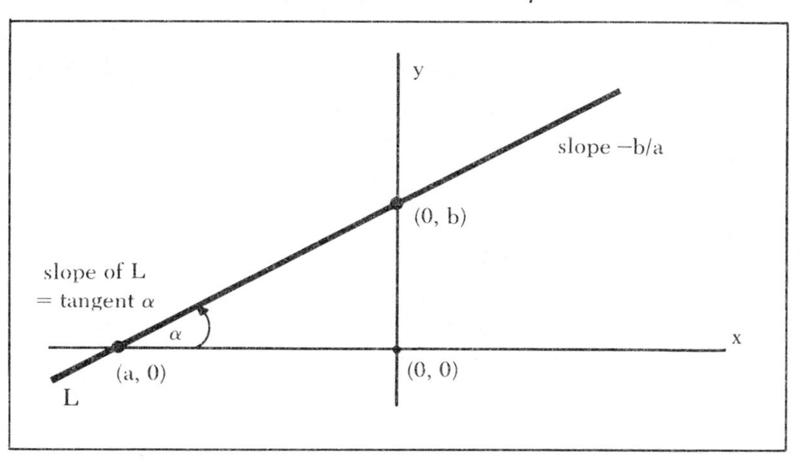

Figure 1–18 The line passing through $(0, b)$ and $(a, 0)$ has slope $-b/a$. The slope of the line is also the tangent of the angle α which the line makes with the x-axis. The angle is measured counterclockwise from the positive half of the x-axis.

Suppose, in particular, that L is a line with slope m passing through the point (x_1, y_1). Then, by applying equation (23) with m replacing the expression $(y_1 - y_0)/(x_1 - x_0)$ (which is simply a computation of the slope), we find

$$y - y_1 = m(x - x_1) \tag{24}$$

as the equation of L. We call equation (24) the *point-slope* form of the equation of a straight line.

POSITIVE SLOPE NEGATIVE SLOPE

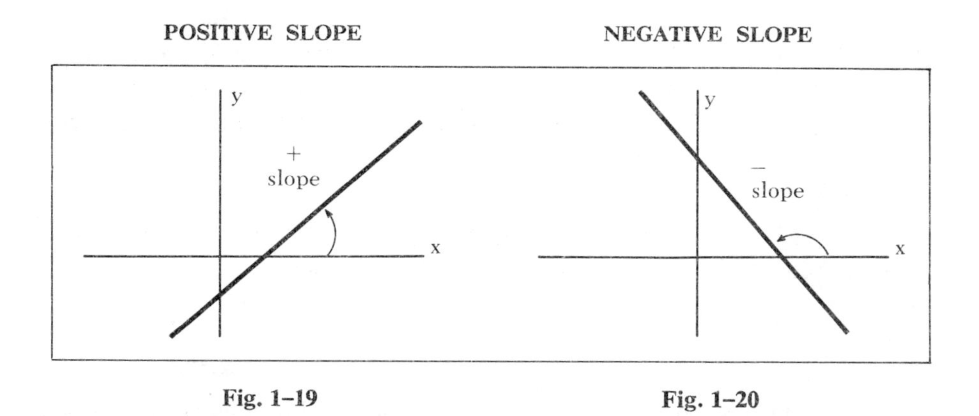

Fig. 1–19 **Fig. 1–20**

Figure 1–19 A line of positive slope makes an acute angle with the x-axis.
Figure 1–20 A line of negative slope makes an obtuse angle with the x-axis.

Example 27: Let L be the line having slope -9 and passing through the point $(-\sqrt{2}, \pi)$. Find the equation of L. Letting $(-\sqrt{2}, \pi)$ be our specific substitution for (x_1, y_1) and $m = -9$, we substitute these values into equation (24). The equation of L becomes

$$y - \pi = -9(x + \sqrt{2}).$$

The line itself appears in Figure 1–21.

LINE WITH SLOPE -9 PASSING THROUGH $(-\sqrt{2}, \pi)$

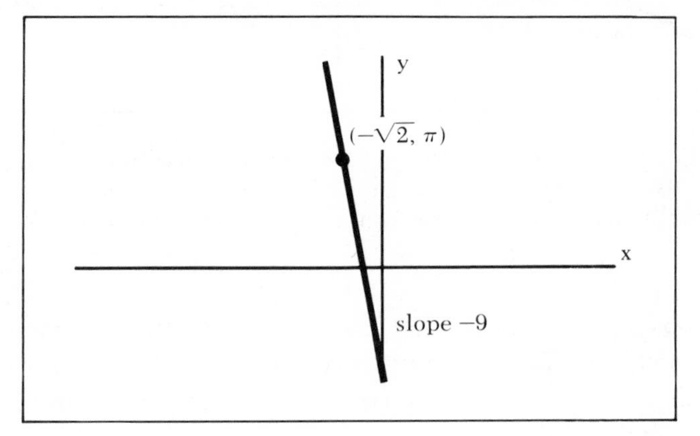

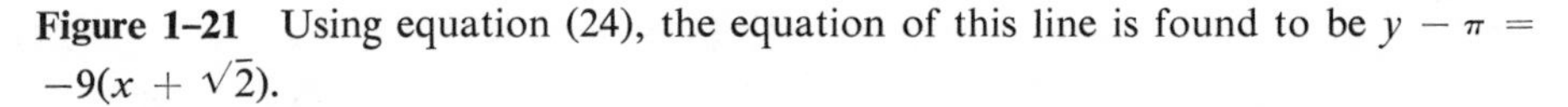

Figure 1–21 Using equation (24), the equation of this line is found to be $y - \pi = -9(x + \sqrt{2})$.

If L is a line with slope m intersecting the y-axis at the point $(0, b)$, then equation (24) is simply

$$y = mx + b. \tag{25}$$

We call equation (25) the *slope-intercept* form of the equation of a straight line. The equation, in any form, of a line that is not parallel to the y-axis can always be reduced by simple algebraic operation to the slope-intercept form.

Example 28: Start with the equation $y - \pi = -9(x + \sqrt{2})$ obtained in Example 27. Then $y - \pi = -9x - 9\sqrt{2}$; hence

$$y = -9x + (-9\sqrt{2} + \pi).$$

From this last equation we may observe that the line has slope -9 and intersects the y-axis at the point $(0, -9\sqrt{2} + \pi)$.

Our problem so far has been to find an equation for L, a straight line in the coordinate plane. As we have seen, the equation for L can always be expressed in the form $x = a$ if L is parallel to the y-axis, or in the form $y = mx + b$ if L is *not* parallel to the y-axis. We first designated a certain type of subset of the coordinate plane and then found an algebraic expression describing that subset.

However, we are often given an equation—for example, $y = x^2 - 1$—and are asked to find the set of points in the coordinate plane whose coordinates (x, y) satisfy the given equation. In the plane, the subset whose points satisfy the given equation is called the *graph* of the equation. The graph enables us to construct a pictorial representation of the equation itself. From such an illustration we can see at a glance many of the properties of the equation and the relationship between the two variables.

Example 29: A corporation decides that a blend of two chocolates should cost \$0.04 per item to produce. Chocolate A and chocolate B cost the corporation \$0.01 and \$0.015 per ounce, respectively; the production costs of each item will be \$0.01 regardless of the item's weight. Let x and y be the amounts (in ounces) of chocolates A and B used in the production of each item. To meet the stated conditions, the corporation's product must then satisfy the equation

$$0.01x + 0.015y + 0.01 = 0.04. \tag{26}$$

We can reduce equation (26) to

$$y = (-2/3)x + 2, \tag{27}$$

which resembles the slope-intercept form of the equation of a straight line (the line with slope $-2/3$, intersecting the y-axis at $(0, 2)$). Note, however, that there can be no negative values for either x or y, since the least amount of either kind of chocolate that one can put into the mix is none. Figure 1–22 indicates all of the points (x, y) in the plane such that

THE CONDITIONS OF EXAMPLE 29

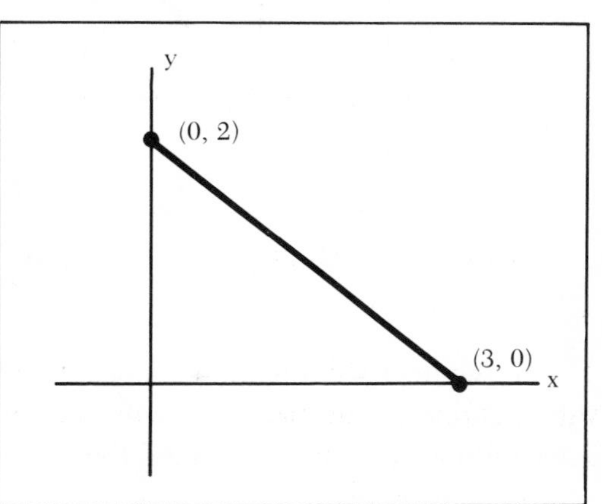

Figure 1–22 If x and y represent the amounts of chocolates A and B, then x and y must satisfy $x \geq 0$, $y \geq 0$, and

$$0.01x + 0.015y + 0.01 = 0.04.$$

x and y satisfy equation (26) and are consistent with the physical limitations of the problem.

The company must then determine the "optimal" choices of x and y to maximize the sales of their product. For example, they may consider such questions as: Is more weight, and hence more A and less B, more important than better taste (more B and less A)? Answers to such questions might then determine what proportions of chocolates A and B should be used.

Example 30: The area A of a circle of radius r is πr^2:

$$A = \pi r^2. \tag{28}$$

Since r is the radius of a circle, it must be non-negative. Clearly, as r increases so does A. Equation (28) involves the two variables A and r; however, the variable A is dependent on r in the sense that once r is selected there is no choice as to what A will be. In other words, A is a function with variable r. We can graph equation (28) in the coordinate plane, using the y-axis for the A values and the x-axis for the values of r. The graph will consist of all points (x, y) such that $y = \pi x^2$. This is shown in Figure 1–23.

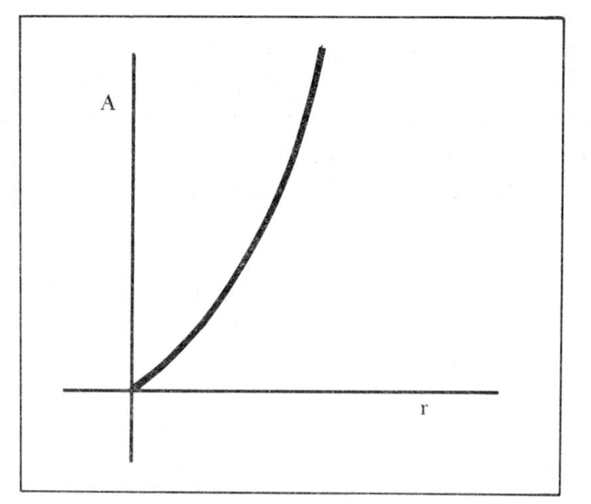

Figure 1–23 The graph of $A = \pi r^2$ is the same as the graph of $y = \pi x^2$.

In the next section we will consider certain types of graphs. Later we will examine techniques that will greatly facilitate sketching graphs and determining basic properties of the equations that the graphs represent.

Exercises

ROUTINE

1. i) Write the equation for each of the straight lines described below. Use any suitable form for your equations.
   ii) Sketch each of the lines.
   iii) Find the slope of each line for which the slope is not explicitly stated.
      a) the line that passes through $(1, 1)$ and $(5, 6)$
      b) the line that passes through $(-1, 4)$ and $(1, -1/4)$
      c) the line with slope -5 passing through $(0, 0)$
      d) the line with slope $\sqrt{2}$ passing through $(-1/2, 1/2)$
      e) the line with slope 6 that intersects the x-axis at $(5, 0)$
      f) the line that intersects the axes at $(0, 1)$ and $(-4, 0)$
      g) the line with slope $6/7$ that intersects the y-axis at $(0, -0.89)$

2. i) Sketch the lines whose equations are given below.
   ii) Express these equations in slope-intercept form in those cases in which the equation does not already have that form.
      a) $y = 3x - 1$
      b) $3x - 4y + 9 = 0$
      c) $y - 1 = 10(x - 5)$
      d) $-y = 4x + 9$

CHALLENGING

3. A journey of 10 miles is to be made along a river. It is possible to paddle a canoe at 6 miles per hour or to carry it along the shore at 3 miles per hour. Let x and y be the number of hours paddling the canoe and carrying it.
   a) Let D be the total distance traveled, both carrying and paddling; express D in terms of x and y.
   b) Sketch a graph illustrating the various possible combinations of x and y for the 10-mile trip.
   c) If it is decided that the trip should take exactly 2 hours, what must be the values of x and y? We can obtain the answer to this question by determining where the graph found in (b) intersects the graph of $x + y = 2$. Explain why.
   d) Suppose that the trip can take any amount of time up to 2 hours. Represent graphically the various possible combinations of x and y.

4. Find the points of intersection of the pairs of lines whose equations appear below. If the two given lines do not intersect, write PARALLEL. Graph each pair of lines on one set of axes.
   a) $x = 3$ and $y = 4x - 1$
   b) $3x - y = 1$ and $x + y = 1$
   c) $x = 3$ and $y = -5$
   d) $y - 3 = 4(x + 6)$ and $y = -7x + 1$
   e) $y = 4x + 2$ and $y - 5 = 4(x - 1/2)$

5. The cost of an item influences the number of sales in a manner that

can be approximated linearly. Given these prices and the corresponding sales, estimate the slope of the line. (Use equation (23) to find the slope.) Estimate what the sales would be if the price were $0.31.

($0.23, 2000), ($0.26, 1840), ($0.29, 1685)

6. Graph each of the following equations for x in the open interval $(-3, 3)$.

a) $y = x$
b) $y = x^2$
c) $y = x^3$
d) $y = x^4$
e) $y = x^5$
f) $y = x^6$

1.5 GRAPHS OF INEQUALITIES. GRAPHS OF FUNCTIONS

Graphing Inequalities of the Form $ax + by < c$

Consider the inequality

$$4x - 3y < 7. \tag{29}$$

The graph of the equation

$$4x - 3y = 7$$

is a straight line. This straight line divides the coordinate plane into two parts separated by the line. The set of points on one side of the line with equation $4x - 3y = 7$ form the solution set to inequality (29), while the points on the other side of the line satisfy the inequality

$$4x - 3y > 7.$$

as shown in Figure 1–24.

We can determine which side of the line satisfies which inequality by testing one point from one side. Since the point (0, 0) represents a solution for inequality (29), all points on the same side as (0, 0) and only those points represent solutions to (29).

To graph an inequality of the form

$$ax + by < c,$$

where a, b, and c are constants, follow these steps:

(1) Graph the straight line $ax + by = c$.

(2) Take a point (x_1, y_1) on some side of the line graphed in Step 1, and note whether $ax_1 + by_1$ is less than or greater than c.

(3) If $ax_1 + by_1 < c$, then all points on the same side of the line as (x_1, y_1) also satisfy the inequality $ax + by < c$, and the points on the side opposite that of (x_1, y_1) will satisfy $ax + by > c$. If $ax_1 + by_1 > c$, then all points on the same side of the line as (x_1, y_1) will

DIVIDING THE PLANE INTO TWO PARTS

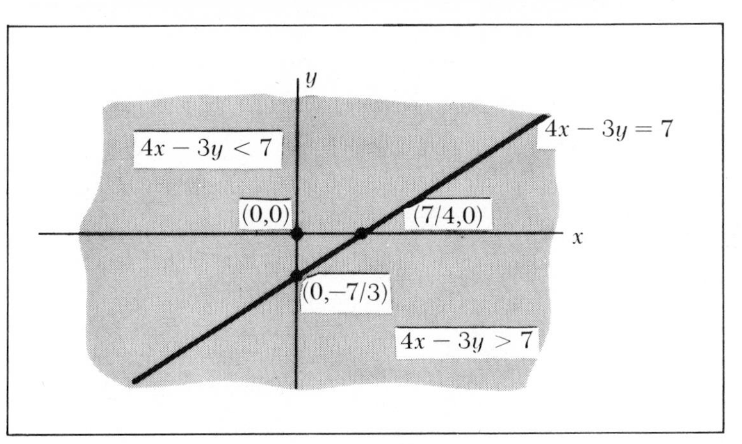

Figure 1–24 The line with equation $4x - 3y = 7$ divides the plane into two parts. The points on one side of the line satisfy the inequality $4x - 3y < 7$; the points on the other side of the line satisfy the inequality $4x - 3y > 7$.

satisfy the inequality $ax + by > c$ and the points on the side opposite that of (x_1, y_1) will satisfy $ax + by < c$.

(4) To find the set of points which satisfy

$$ax + by \leq c,$$

we find the points which satisfy $ax + by < c$, and then add the points *on the line* with equation $ax + by = c$. This is the complete solution set for $ax + by \leq c$.

Example 31: Find the solution set to the simultaneous inequalities

$$2x + 3y < 1$$
$$y \geq 2x - 1.$$

In Figure 1–25 we graph the points which satisfy $2x + 3y < 1$; the points which satisfy $y \geqslant 2x - 1$ are graphed in Figure 1–26. Note that in Figure 1–26 the line itself forms part of the solution set, since we have $\geqslant$ rather than the strict inequality $>$. The solution set to the simultaneous inequalities is the intersection of their solution sets. This intersection is graphed in Figure 1–27.

Graphs of Functions

Thus far we have only considered graphs for equations and inequalities. It is a very short step, however, from the graph of an equation

SOLUTION SET FOR $2x + 3y < 1$

Figure 1–25 The solution set does not include the points on the line with equation $2x + 3y = 1$.

SOLUTION SET FOR $y \geq 2x - 1$

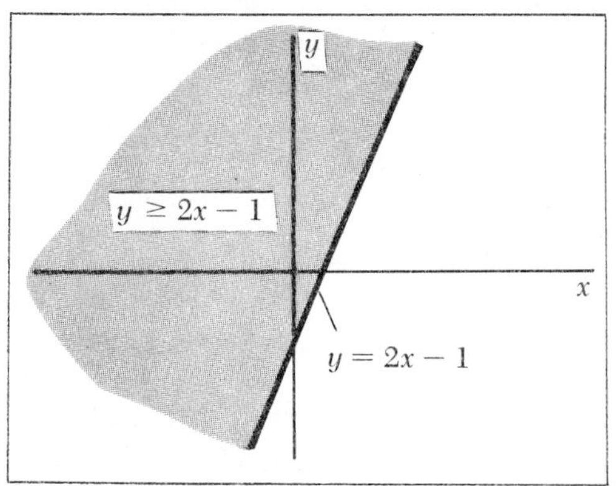

Figure 1–26 The solution set includes the points on the line with equation $y = 2x - 1$.

SOLUTION SET FOR THE SIMULTANEOUS INEQUALITIES $2x + 3y < 1$ AND $y \geq 2x - 1$

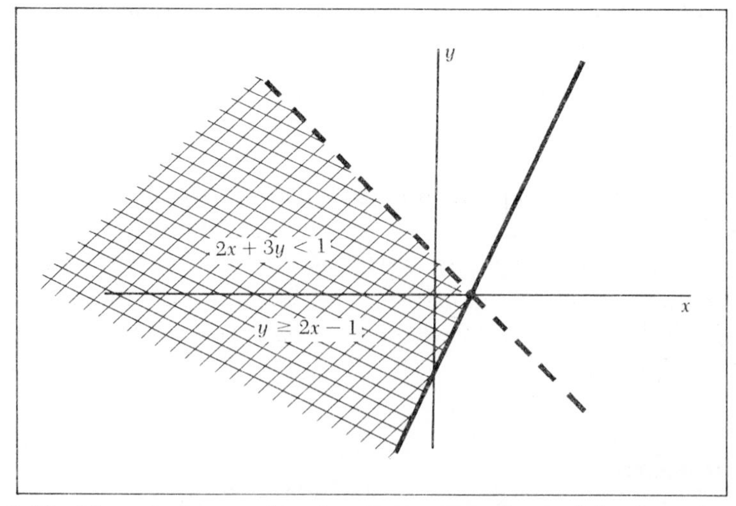

Figure 1–27 The solution set for the simultaneous inequalities $2x + 3y < 1$ and $y \geq 2x - 1$ is found by taking the intersection of the solution sets graphed in Figures 1–25 and 1–26.

to the graph of a function. The graph of a function is simply the graph of the equation which defines the function.

Example 32: Let

$$f(x) = \begin{cases} x, & \text{if } x \geq 0, \\ -x, & \text{if } x < 0. \end{cases} \tag{30}$$

Here the definition of $f(x)$ holds for any real number x; hence the domain of f is assumed to be R. Note also that the definition of $f(x)$ consists of two parts. Given any x, we must observe whether x is less than, or greater than or equal to, 0. Then we apply the appropriate part of equation (30) to evaluate $f(x)$. Even though the definition of $f(x)$ is divided into two parts, $f(x)$ is uniquely determined for any specific value of x; therefore equation (30) does in fact define a function. Actually, the function can be characterized in a more compact form by the equation $f(x) = |x|$.

If we think of $f(x)$ as a variable whose value is dependent on x, then—as in Example 30—the equation involves two variables and can be graphed. We determine the set of points (x, y) in the coordinate plane such that $y = f(x)$; in this case $y = |x|$. For $x \geqslant 0$, the graph coincides with the line whose equation is $y = x$; for $x < 0$, the graph is the line whose equation is $y = -x$ (Figure 1–28).

GRAPH OF THE FUNCTION f

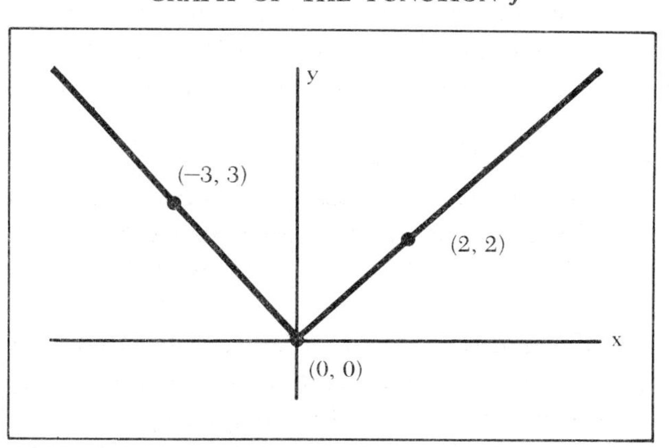

Figure 1–28 The graph of the absolute value function (defined in equation (30)) is the graph of $y = f(x) = |x|$.

Definition 8: *The **graph** of a function f is the graph of the equation $y = f(x)$ for all x for which $f(x)$ is defined.*

We pause for a moment to review the logical sequence begun in Section 1.4. Starting with straight lines in the plane, we discovered that there are equations in the variables x and y that characterize straight lines. Next we associated with each equation in two variables a subset of the coordinate plane; we called this subset the graph of the equation. The graph of an equation is simply the set of points (x, y) of the plane such that x and y satisfy the given equation. Now, given any function f, we associate with f the equation $y = f(x)$, an equation in two variables, and call the graph of this equation the graph of f. The graph of f thus consists of all points in the plane of the form $(x, f(x))$.

Example 33: A ball is dropped from a height of 1600 feet. The speed at which the ball is traveling after t seconds is

$$s(t) = 32t \text{ (feet per second)} \tag{31}$$

and the distance the ball has traveled after t seconds is

$$d(t) = 16t^2 \text{ (feet).} \tag{32}$$

We make no attempt to justify equations (31) and (32) at the moment, but we will have the tools to do so by the end of this discussion. Both s and d are functions of t, the time elapsed since the ball was dropped.

We obtain the graphs of s and d by graphing $y = s(t)$ and $y = d(t)$. Here t, instead of x, is used as the function variable, but this variation in labeling is insignificant; we can equally well imagine a t-axis in place of the x-axis.

The graphs of s and d are plotted together in Figure 1–29, We have not allowed negative values for t, because positive time elapses once the ball is released. Nor have we allowed values of t greater than 10, for at $t = 10$ the ball will have traveled 1600 feet, and will have come to rest on the ground.

The graphs of s and d in Figure 1–29 indicate that both the speed and distance the ball has traveled increase as t increases. Examining the graph of s, however, one suspects that s is increasing at a "constant" rate. If we choose some value t_0 for t and consider time $t_0 + h$—that is, t_0 increased by h seconds—then the *increase* in speed during the interval t_0 to $t_0 + h$ is the new speed at $t_0 + h$ minus the old speed at t_0:

$$s(t_0 + h) - s(t_0) = 32(t_0 + h) - 32(t_0) = 32h. \tag{33}$$

This increase of $32h$ is not dependent on the value of t_0, the initial time, but only on the value of h, the time elapsed since t_0. Thus, *the rate of*

GRAPHS OF SPEED AND DISTANCE OF A FALLING BALL

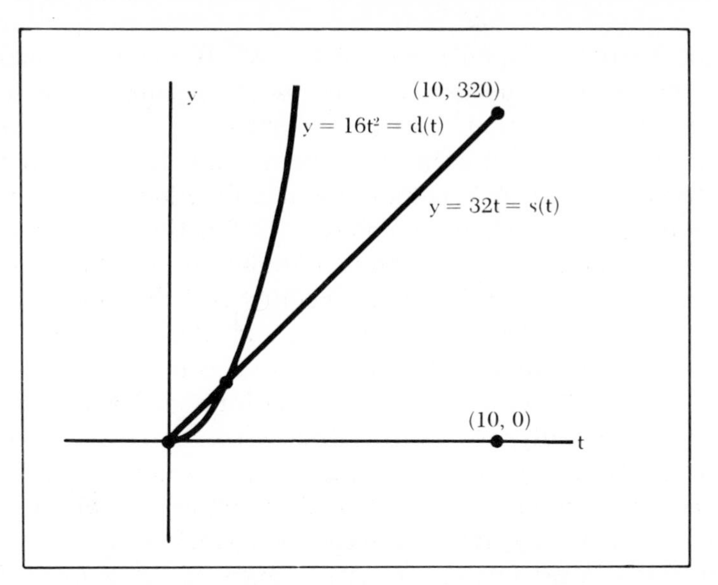

Figure 1–29 The graphs of the speed and distance traveled by the ball are shown on the same set of coordinates. The ball has traveled $16t^2$ feet and is moving at $32t$ feet per second at t seconds after it has been dropped.

increase of the speed is

$$\frac{\text{increase in speed}}{\text{time for increase}} = \frac{32h}{h} = 32, \tag{34}$$

which is a constant.

It is also clear from the graph of d that $d(t)$ is *not* increasing at a constant rate; indeed, $d(t)$ seems to increase more and more rapidly as t increases. Applying the method used for s, we see that the increase in distance during the time interval from t_0 to $t_0 + h$ is

$$d(t_0 + h) - d(t_0) = 16(t_0 + h)^2 - 16(t_0)^2 = 32t_0h + 16h^2. \tag{35}$$

Now the increase depends not merely on h, but also on t_0. The larger t_0 is, the larger is the increase in distance during the interval h. These facts confirm our suspicion that $d(t)$ increases more rapidly the larger t becomes. The rate of increase of $d(t)$ over a time interval of length h is expressed as

$$\frac{\text{increase in distance}}{\text{time for increase}} = \frac{32t_0h + 16h^2}{h} = 32t_0 + 16h, \tag{36}$$

and this rate depends on both h and t_0.

We will return to this central notion of rate of increase later on.

With regard to graphing, one might ask: It may sometimes be true that the graph of a function enables us to visualize some important properties of the function, but isn't it necessary to know these same properties before we can even draw the graph? If we did not have a substantial amount of information about a function, how could we be sure that any graph we claim to represent the function is in fact accurate?

First, one should keep in mind that graphs, like most pictures used in mathematics, serve primarily to aid memory and the rapid assimilation of data. Prior to constructing an accurate and adequate picture or graph, one needs a certain amount of information. It is this information, rather than the graph itself, that is of real mathematical significance and practical importance. Secondly, we will acquire certain tools that will greatly facilitate graphing and the investigation of functions. Until some of these tools are presented, we will simply show graphs without detailed discussion of how they were obtained. Nevertheless, in many instances one can obtain a moderately accurate sketch of the graphs of certain functions simply by inspecting the definitions of the functions, plotting a few selected points, and using mathematical "common sense." We conclude this section with an example illustrating how a graph is found by inspection of the function.

Example 34: Consider the function defined by

$$f(x) = 1/(x^2 + 1). \tag{37}$$

Note first that since x^2 is always non-negative, $x^2 + 1$ is at least 1 for any value of x. Consequently, $1/(x^2 + 1)$ is at most 1 for any value of x, and $f(x)$ is defined for every real number x. At the same time, since $x^2 + 1$ is positive, $f(x)$ is also positive for all values of x. It follows that

$$0 < f(x) \leq 1$$

for all values of x. The graph of f will therefore lie entirely between the lines whose equations are $y = 0$ and $y = 1$. The least value that $x^2 + 1$ assumes is 1; this occurs when $x = 0$. Therefore, the largest value of $f(x)$ is $1/1 = 1$, at $x = 0$. If x is positive and increasing, $f(x)$ is decreasing. If x is negative and decreasing, $f(x)$ is also decreasing. We can also conclude that since $f(x) = f(-x)$, the portion of the graph to the left of the y-axis will be the mirror image of the portion to the right of the y-axis (Figure 1–30). We collect this information to obtain an approximation to the graph of f; Figure 1–31 enables us to recall at a glance much of the information derived about f.

SYMMETRY WITH RESPECT TO THE y-AXIS

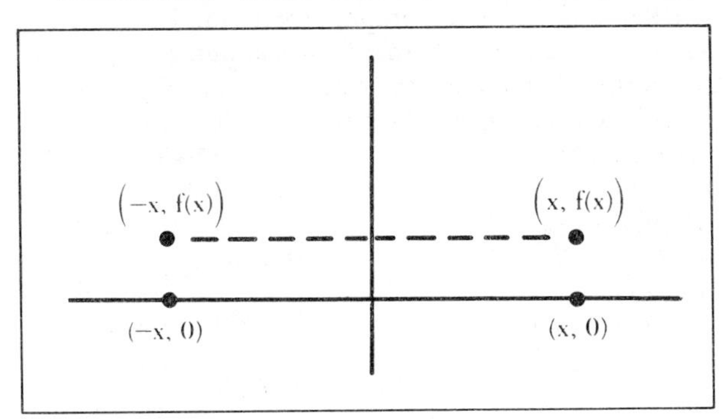

Figure 1–30 Since for any real number x we have $f(x) = f(-x)$, the points $(x, f(x))$ and $(-x, f(x))$ are both on the graph of f; $(-x, f(x))$ is the mirror image of $(x, f(x))$.

A FIRST APPROXIMATION TO THE GRAPH OF $f(x) = 1/(x^2 + 1)$

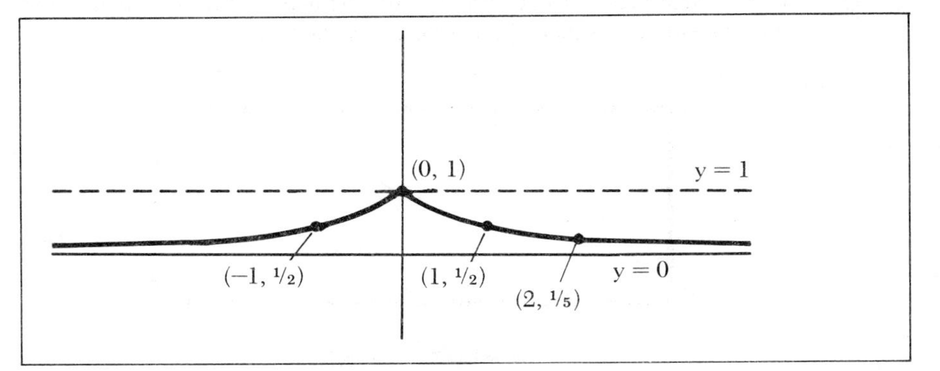

Figure 1–31 The information gathered about f is pictorially represented in this figure.

Exercises

ROUTINE

1. Sketch the solution sets to each of the following inequalities.
   a) $y < x$
   b) $y > x$
   c) $y \leq x$
   d) $2y < 3x - 1$
   e) $2x - y \leq 1$
   f) $10x - 13y - 7 > 0$
   g) $x/4 - y/5 \geq 1$
   h) $-x - y - 1 < 9$

2. Sketch the solution set to the simultaneous inequalities in each of the following.
   a) $y < x$
      $2x < 1$
   b) $2x - y \leq 1$
      $3x + 2y > 8$
   c) $y > 0$
      $2x - y \leq 5$
      $x/2 - y/3 < 1$

3. Sketch the graph of each of the following functions. You should find certain key points on the graph; for example, determine the function values at $x = 0$ and $x = 1$, the set of real numbers for which the function is defined (this determines the "extent" of the graph), and, in these simple instances, the intervals in which the graph is "rising" and "falling."

a) $f(x) = x$
b) $f(x) = -3x$
c) $g(x) = 4x - 5$
d) $h(x) = 5$
e) $f(x) = x^2$
f) $h(x) = -x^2 - 1$
g) $f(x) = x^3$
h) $f(x) = 1/x$
i) $f(x) = \begin{cases} -3x + 1, & \text{if } x \geq 1 \\ 2x, & \text{if } x < 1 \end{cases}$
j) $h(x) = 1/x^2$
k) $f(x) = \begin{cases} 1, & \text{if } x \neq 0 \\ 0, & \text{if } x = 0 \end{cases}$

CHALLENGING

4. A man has a total of 100 feet of fencing. He uses the fencing to enclose a rectangular area of length x and width y (Figure 1–32). Let $A(x)$ be the area (measured in square feet) enclosed by the fence.

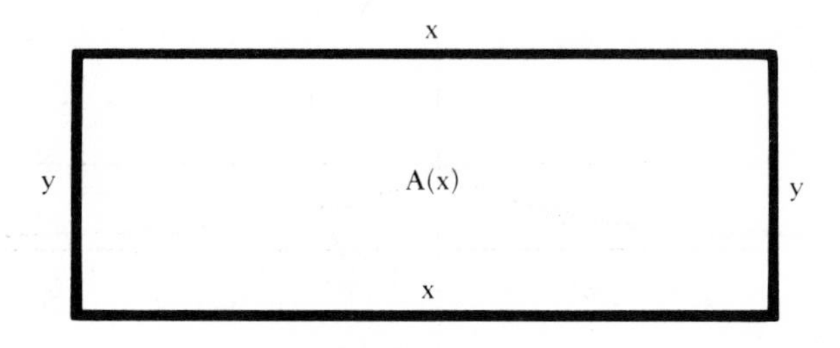

Figure 1–32 Exercise 4.

a) Find $A(x)$ explicitly in terms of x. [Hint: Since the fencing totals 100 feet, the perimeter is $2x + 2y = 100$. The enclosed area is $A(x) = xy$. Solve the first equation for y in terms of x and make the appropriate substitution in $A(x)$.]
b) Graph $A(x)$.
c) Approximate that value of x for which $A(x)$ will be a maximum. (Methods will be introduced later which will enable us to find this maximum value exactly.)

5. What is the average rate of change of velocity of a car going 10 m.p.h. at 10:25 and 50 m.p.h. at 10:35?

6. What is the average rate of change of growth of a one-year-old child who grows 1/4 inch per month and a two-year old child who grows 1/8 inch per month?

7. If a bug goes straight toward first base from home plate at 7:13 and arrives there at 9:43 after a trip of 90 feet, what is his average rate of change of position?

THEORETICAL

8. Graph each of the following equations. In each group graph as many as possible on the same set of axes.

a) $y = -12x^2$
$y = -3x^2$
$y = -2x^2$
$y = -x^2/2$
$y = 0x^2$
$y = x^2/2$
$y = 2x^2$
$y = 3x^2$
$y = 12x^2$

b) $y = x^2 + 0$
$y = x^2 + 3$
$y = x^2 - 5$

c) $y = (x - 0)^2$
$y = (x - 3)^2$
$y = (x + 4)^2$

d) $y = 3(x + 4)^2 - 5$
$y = (-1/2)(x - 2)^2 + 1$

Review of Chapter 1

A *set* is any well-determined collection of objects. The objects comprising a set are called *elements* of the set. Given sets S and T, we can form $S \cup T$ and $S \cap T$.

A set S is a *subset* of a set T if every element of S is also an element of T. Special subsets of real numbers are *open* and *closed* intervals.

We denote the distance between two real numbers a and b by $|a - b|$. If $p > 0$, then the set of points x such that $|a - x| < p$ is the same as the open interval $(a - p, a + p)$.

The set of points (x, y) of the coordinate plane satisfying any given equation $y = f(x)$ form the *graph* of the equation. A *function* f defined on a set S of real numbers is a rule or relationship that assigns a unique real number $f(x)$, called the *function value*, to each element x of S. If f is a function defined on a set S, then the graph of the equation $y = f(x)$ for all x in S is called the *graph* of f.

The graph of an equation of the form $y = mx + b$ is a straight line with slope m. The equation of the straight line passing through the points (x_0, y_0) and (x_1, y_1) of the coordinate plane is

$$y - y_0 = \left(\frac{y_1 - y_0}{x_1 - x_0}\right)(x - x_0),$$

if $x_1 \neq x_0$, and $x = x_0$, if $x_1 = x_0$.

In addition to graphing equations and functions, it is possible to graph the solution sets to inequalities involving one or two variables.

REVIEW EXERCISES

1. Sketch each of the following sets on a line representing the real numbers.

a) $\{x \mid 2 \leq x \leq 4\}$
b) $\{x \mid 2 \leq x\}$
c) $\{x \mid x^2 < 4\}$
d) $\{x \mid |x| < 2\}$
e) $\{x \mid x < 2\} \cap \{x \mid x > -3\}$
f) $\{x \mid x < 2\} \cup \{x \mid x > -3\}$
g) $\{x \mid x^2 - 1 > 0\}$
h) $\{w \mid |w - 3| < 0.5\}$
i) $\{z \mid |3 - z| < 0.5\}$
j) $\{x \mid 2 < x < 2 \cdot 7\} \cap \{w \mid -1 \leq w \leq 4\}$
k) $(\{x \mid x > 3 \cdot 4\} \cap \{y \mid y \leq 5 \cdot 6\}) \cup \{w \mid w > -5\}$
l) $\{x \mid |x - 1| < 4\} \cap \{w \mid 2 \leq w < 4\}$

2. A merchant wishes to determine the price per unit a customer should pay if he buys any number from 1 to 100 of a certain item. If one of the items can be purchased for \$2, while the items sell for \$1 each in quantities of 100, find the straight line equation the merchant uses to determine the cost of n items; $n = 1, \ldots, 100$. What is the cost per unit when a customer purchases 65 items?

3. A factory figures the profit it will make in selling n units of a certain item according to the function

$$p(n) = 100n - n^2 = -n(n - 100).$$

Find the profit realized from selling 10 items, 90 items, 60 items, and 40 items. After how many items will the factory begin to lose money? Estimate the number of items the factory must sell to realize maximum profit. Graph the function p.

4. Graph each of the following equations or functions.

a) $y = x$
b) $y = -x$
c) $f(x) = x$
d) $y = x^2$
e) $y = 3x - 8$
f) $y = -7x + 0.5$
g) $f(x) = -6x - 10$
h) $f(x) = 5$
j) $y = 1/(x^2 + 1)$
k) $f(x) = 3(x - 1) + 4$

5. Graph the solution sets for the following inequalities.

a) $y < x + 1$
b) $2y < x + 1$
c) $2y \geq 3x + 1$
d) $x/2 + y/2 < -1$

Vocabulary

sequence
limit
convergence
continuity

Chapter 2

Limits and Continuity

2.1 LIMITS

Since $\sqrt{2}$ is an irrational number, it is impossible to obtain an exact decimal representation of it. We can, using the algorithm for computing square roots, find a decimal representation of $\sqrt{2}$ accurate to as many decimal places as desired. Increasingly accurate decimal representations of $\sqrt{2}$ are:

$$\begin{array}{l} 1 \\ 1.4 \\ 1.41 \\ 1.414 \\ 1.4142 \\ \cdot \\ \cdot \\ \cdot \end{array}$$

We could extend this list indefinitely. Moreover, the further we extend the list, the better approximation of $\sqrt{2}$ we obtain.

No ending decimal number exactly equals $\sqrt{2}$. If, however, we have a particular degree of approximation in mind, then the desired accuracy will be obtained at some number on the list and will be maintained for each number thereafter.

Another way of expressing the above idea is the following: Choose any positive number p. Then, as we saw in Chapter 1, the open interval $(\sqrt{2} - p, \sqrt{2} + p)$ consists of those numbers that approximate $\sqrt{2}$ with accuracy p: those numbers x such that $|x - \sqrt{2}| < p$. The list given above, if extended indefinitely, will eventually include some point of $(\sqrt{2} - p, \sqrt{2} + p)$, regardless of how small p is. Moreover, once some decimal approximates $\sqrt{2}$ with accuracy p, all subsequent decimals in the list will approximate $\sqrt{2}$ with at least the same accuracy, and hence will also be in $(\sqrt{2} - p, \sqrt{2} + p)$. We conclude that, regardless of how small a positive number p we choose, the list will from some point on consist entirely of members of $(\sqrt{2} - p, \sqrt{2} + p)$ (Figure 2–1). Because of the properties of the list with regard to $\sqrt{2}$, it is not unreasonable to say that the terms of the list approach $\sqrt{2}$ as a *limit.*

DECIMAL APPROXIMATION TO $\sqrt{2}$

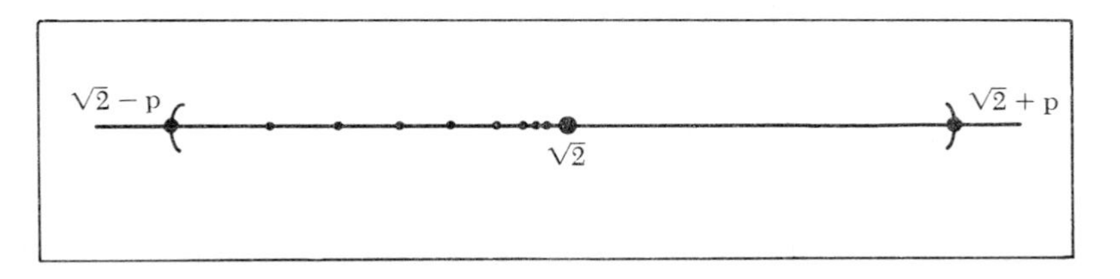

Figure 2–1 For any $p > 0$, a decimal approximation to $\sqrt{2}$ can be found in the interval $(\sqrt{2} - p, \sqrt{2} + p)$. If some decimal approximation in the list is inside the interval, then all subsequent approximations will also be inside the interval.

We momentarily leave our considerations concerning approximations of $\sqrt{2}$ and consider the extended Example 33 of Chapter 1. In equation (36) of that example, we saw that the rate of increase of distance during the interval from time t_0 to time $t_0 + h$ is given by

$$\frac{\text{increase in distance}}{\text{time for increase}} = 32t_0 + 16h. \tag{1}$$

More precisely, this is the *average speed* in the time interval from t_0 to $t_0 + h$. In other words, this is simply

$$\frac{\text{total distance traveled from } t_0 \text{ to } t_0 + h}{\text{total time from } t_0 \text{ to } t_0 + h\ (= h)}.$$

The *rate of increase in distance* is simply another term for *velocity* or *speed;* for example, if a car is traveling at 30 miles per hour, then the distance traveled is increasing at a rate of 30 miles per hour. Therefore, equation (1) is a measure of the speed in the time interval from t_0 to $t_0 + h$.

We generally associate speed with a particular moment in time; for example, we think of a car's speed when we observe the speedometer. We can, of course, talk about average speed over the course of a trip, but we usually talk of the speed of a car with regard to particular moments during a trip.

If we choose $h = 0$, and therefore let $t_0 = t_0 + h$, then the left side of equation (1) becomes undefined. Explicitly, when $h = 0$, both the increase in distance and the time for the increase are zero; the left side of equation (1) then has the meaningless form 0/0. Yet it is possible to substitute $h = 0$ into the right side of equation (1) without logical inconsistency, and it becomes simply $32t_0$. If we refer to Example 33 of Chapter 1, we see that $32t_0$ is in fact the actual speed of the ball at time t_0. This, as we will see later, is not a coincidence.

While choosing $h = 0$ in equation (1) leads to a meaningless expression on the left, the expression

$$\frac{\text{increase in distance}}{\text{time for increase}} \tag{2}$$

retains physical significance for any value of $h > 0$, and can readily be computed using the right side of equation (1). We can therefore sensibly discuss what happens to expression (2) as we choose smaller and smaller values for h (as long as these values are not zero). Clearly, as one chooses values for h closer and closer to zero, expression (2) becomes a better and better approximation to $32t_0$, the speed at t_0.

Although this example of velocity is quite different in character from the approximation of the irrational number $\sqrt{2}$, the notion of limit enters both in almost the same way. In this case, expression (2) approximates $32t_0$ with any degree of accuracy we wish, as long as h is chosen small enough. In particular, if we want expression (2) to approximate $32t_0$ with accuracy $p > 0$, then since (2) is equal to $32t_0 + h$ for any $h > 0$ and

$$|(32t_0 + h) - 32t_0| = h,$$

we need $0 < h < p$ to get the desired degree of accuracy.

In the following definitions, we try to make more precise certain notions introduced so far in this chapter.

Definition 1: *A* ***sequence*** *is a list of numbers containing a first element, a second element, and so on, with one element for each positive integer.*

More formally, a sequence is a function from the set of positive integers into the real numbers; that is, an assignment of a real number to each positive integer. Although this formal definition of sequences is more precise from a mathematical point of view, the less formal definition given above reflects the manner in which sequences are often thought of and discussed from a practical point of view.

Example 1: The extended list given at the beginning of this chapter is a sequence with first term 1, second term 1.4, and so on. The nth term of this sequence, the number occurring in the nth place in the list, is the decimal representation of $\sqrt{2}$ correct to n decimal places.

Definition 2: *An element of a sequence is called a* ***term*** *of the sequence. For any positive integer n, the element of the sequence occupying the* n^{th} *place is called the* $\mathbf{n}^{\text{th}}$ ***term.*** *If we use some letter such as s to denote a sequence, it is customary to use* s_n *to denote the* n^{th} *term. We may also denote this sequence by writing*

$$s_1, s_2, s_3, s_4, \ldots$$

Example 2: Consider the sequence determined by letting $s_n = n^2$. Note that the sequence is specified when we state precisely what its nth term will be for any positive integer n. The first five terms of this sequence are

$$1, 4, 9, 16, 25.$$

There is no reason why all terms of a sequence need be distinct, as the following example shows.

Example 3: Set $s_n = 3$ for every positive integer n. In the sequence s thus defined, each of the terms is 3.

Definition 3: *We say that a sequence has a* ***limit*** *L, or that s converges to L, if the terms of s eventually approximate L to any desired degree of accuracy p, where* $p > 0$.

More formally, s converges to L if, given any $p > 0$, it is possible to find a positive integer M, which usually depends on p, such that $|s_n - L| < p$ for all $n \geq M$. Again, this means that once we have chosen a degree of accuracy, $p > 0$, all of the terms of the sequence from some term on will approximate L with accuracy p.

Example 4: The sequence given at the beginning of this chapter converges to $\sqrt{2}$.

Example 5: The sequence in Example 2 has no limit because its terms do not tend to approximate any real number.

Example 6: The sequence of Example 3 converges to 3, because all of its terms are perfect approximations to 3; that is, they equal 3.

Example 7: The sequence defined by

$$s_n = 1 - (1/n)$$

converges to 1. This convergence is illustrated in Figure 2–2. As n increases, the points $(n, s_n) = (n, 1 - (1/n))$ approach the line $y = 1$. The y-coordinate of (n, s_n) converges to 1 as n is chosen larger and larger.

CONVERGENCE OF A SEQUENCE

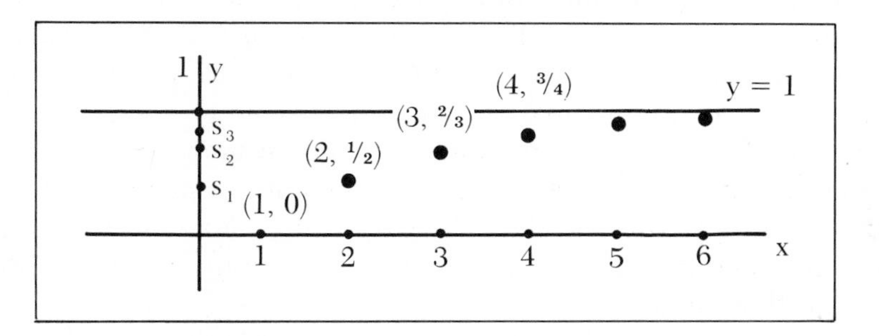

Figure 2–2 We graph the sequence by representing the term $s_n = 1 - (1/n)$ by the point $(n, 1 - (1/n))$. The convergence of the sequence to 1 is illustrated by the graph approaching the line $y = 1$ as n increases.

Having defined the limit of a sequence, we now explore the concept of the limit of a function.

Definition 4: *A function f has a* ***limit*** *L as x approaches a if (i) f(x) is defined for all points x other than a in some open interval which contains a; and (ii) f(x) will be as close as we wish to L provided x is chosen in the domain of f sufficiently close to, but not equal to, a (Figure 2–3).*

THE LIMIT OF $f(x)$ AS x APPROACHES a IS L

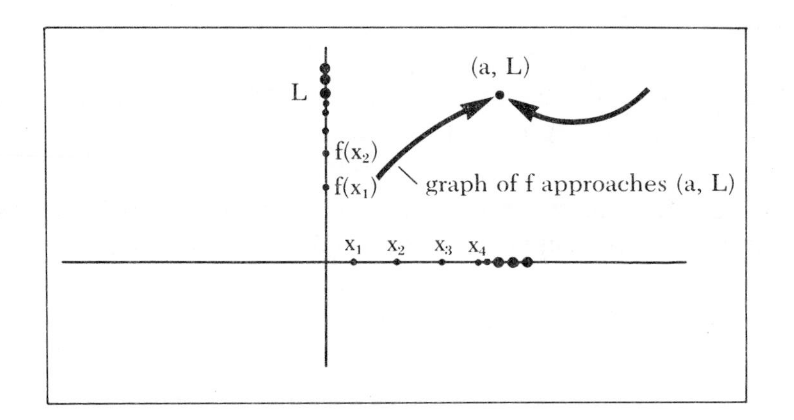

Figure 2–3 If a sequence $x_1, x_2, \ldots$ converges to a, and if the limit of $f(x)$ as x approaches a is L, then the sequence $f(x_1), f(x_2), \ldots$ converges to L.

A more formal statement of requirement (ii) can be stated as follows. Given any $p > 0$, it is possible to find some $q > 0$ such that when $|x - a| < q$ and $x \neq a$, then $|f(x) - L| < p$. The number p is the desired degree of approximation, while q tells us how closely x must approximate a for $f(x)$ to approximate L with accuracy p.

Here is yet another way of defining the limit of a function: Given any $p > 0$, it is possible to find $q > 0$ such that if $x \neq a$, if x is in the domain of f, and if x approximates a with accuracy q, then $f(x)$ approximates L with accuracy p.

Thus, for example, the function in equation (1) has the limit $32t_0$ as h approaches 0. Further examples follow.

Example 8: Let $f(x) = x^2$. Evidently, as x approaches 2, $f(x)$ will approach 4 (Figure 2–4). Thus, in this case, the limit of $f(x)$ as x approaches 2 is $f(2)$.

Example 9: Consider $f(x) = 1/x$. As x approaches 0, $f(x)$ does not approach any limit at all. This is evident from the graph of f in Figure 2–5: as x approaches 0, the parts of the graph on opposite sides of the y-axis do not tend to approach a common point, or any other point for that matter.

The type of limit presented in the following example is of central importance in the calculus.

THE LIMIT OF $f(x) = x^2$ AS x APPROACHES 2 IS 4

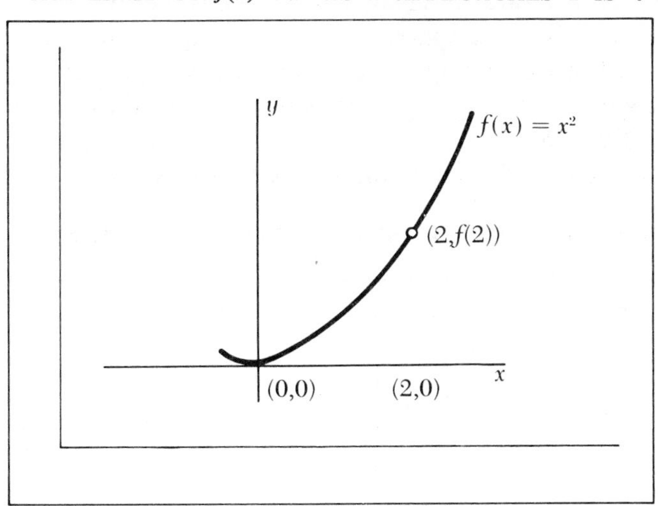

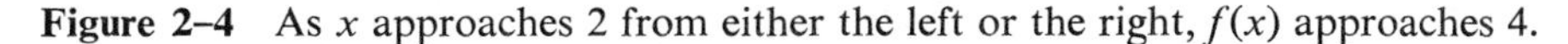

Figure 2–4 As x approaches 2 from either the left or the right, $f(x)$ approaches 4.

THE FUNCTION $f(x) = 1/x$ HAS NO LIMIT AS x APPROACHES 0

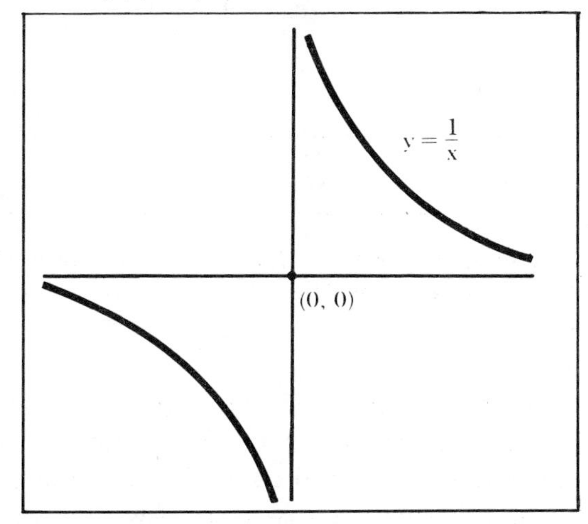

Figure 2–5 As x approaches 0 from the right, $f(x)$ becomes more and more positive. As x approaches 0 from the left, however, $f(x)$ becomes more and more negative.

Example 10: Let $f(x) = x^2 - 1$. Extending the ideas introduced in Example 33 of Chapter 1 and earlier in this chapter, we can reasonably state that the average rate of change of $f(x)$ between 2 and some number $2 + h$ is given by

$$\frac{\text{total change in the function value}}{\text{total change in the variable}} = \frac{f(2 + h) - f(2)}{(2 + h) - 2}$$

$$= \frac{((2 + h)^2 - 1) - (4 - 1)}{h} = \frac{4h + h^2}{h} = 4 + h. \tag{3}$$

Thus, the average rate of change in this instance is a function of h. Although there is no apparent difficulty encountered in substituting $h = 0$ in the last expression in equation (3), when $h = 0$ we again lose meaning for the preceding expressions in (3), since they assume the form $0/0$ on substitution of $h = 0$. As before, however, we can talk about a *limit* as *h approaches* 0; the limit, as one may observe, is 4 (Figure 2–6).

A FUNCTION UNDEFINED AT SOME POINT MAY HAVE A LIMIT THERE

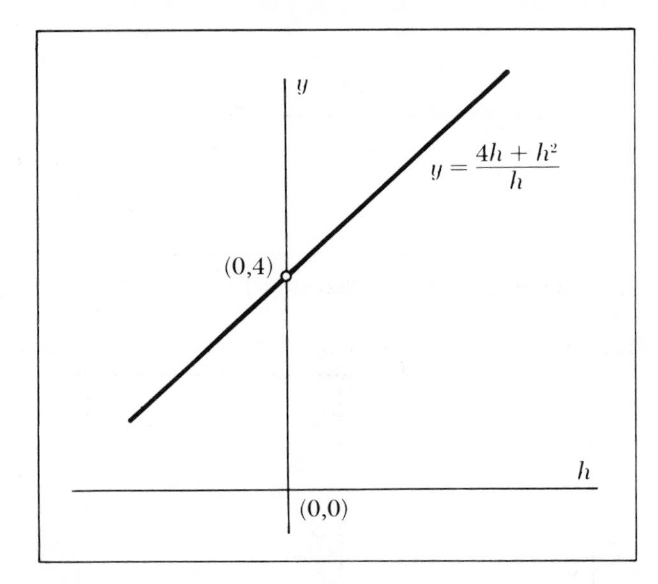

Figure 2–6 Although $(4h + h^2)/h$ is not defined for $h = 0$, it does have 4 as a limit as h approaches 0. We can think of this limit as the rate of change of $f(x) = x^2 - 1$ when $x = 2$.

This limit may be thought of as the rate of change of $f(x)$ when $x = 2$. Several interpretations for the limit of equation (3) as h approaches 0 will be discussed in a more general context later. For the moment, we may think of this limit as what we would see if we looked at the "speedometer" of the function f at the moment when $x = 2$.

Exercises

ROUTINE

1. Write out the first five terms of each of the following sequences. Graph these first five terms in the manner of Figure 2–2.

a) $s_n = 1$ c) $s_n = 1/(n + 1)$
b) $s_n = (-1)^n$ d) $t_n = 1/n^2$

e) $a_n = n$
f) $t_n = n^2 - n + 1$
g) $k_n = n/(n^2 + 1)$ [Hint: Divide numerator and denominator by n^2.]
h) $s_n = (3n - 1)/(5n + 1)$ [Hint: Divide numerator and denominator by n.]
i) $t_n = (5 - 6n)/(7 + 3n)$

2. For each of the functions given below:
   i) Find the limit of the function as the function variable approaches 0. If there is no limit, write *no limit.*
   ii) Graph each of the functions for values of the function variable close to 0.
   iii) Determine whether the limit of the function is the same as the function value at $x = 0$.
      a) $f(x) = 0$
      b) $f(x) = x$
      c) $g(t) = 4t - 5$
      d) $m(t) = (t^2 - 4)/(t - 2)$
      e) $h(t) = (4 - t)/t$
      f) $\dfrac{(-1 + h)^2 - 1}{h} = p(h)$
      g) $f(x) = \begin{cases} 0, & \text{if } x \geq 0, \\ 2, & \text{if } x < 0 \end{cases}$
      h) $m(x) = \dfrac{x^3 - 8}{x - 2}$
      i) $g(x) = 1/x^2$

3. Follow the instructions of Exercise 2 for the functions in Exercise 2 with $x = 2$ substituted for $x = 0$.

CHALLENGING

4. Indicate the rule by which the nth term is computed for each of the following sequences. Graph the first five terms of each sequence in the manner of Figure 2–2.
   a) 2, 2, 2, 2, 2, . . .
   b) 1/2, 1/3, 1/4, 1/5, 1/6, . . .
   c) $2\frac{1}{2}, 2\frac{1}{3}, 2\frac{1}{4}, 2\frac{1}{5}, 2\frac{1}{6}, \ldots$
   d) 3.9, 3.99, 3.999, 3.9999, 3.99999, . . .
   e) 9, 28, 9, 9, 9, 9, . . .
   f) 1/2, 3/2, 2/3, 4/3, 3/4, 5/4, 4/5, 6/5, 5/6, . . .

5. At the beginning of 1900, a depositor had exactly \$100 in a bank which paid 5 per cent interest per year compounded annually. The interest paid at the end of 1900, therefore, was (100)(0.05) = \$5, and the total in the account at the beginning of 1901 was \$105.
   a) Let s_n be the amount in the account n years after the beginning of 1900, assuming that the account is left undisturbed and the bank continues to pay interest of 5 per cent per year. Find s_n for $n = 1, 2, 3, 4,$ and 5.
   b) Show that $s_n = s_{n-1}(1.05)$. Does the sequence s converge?

6. Let $f(x) = x^2$. Each of the following sequences converges to 2. For each sequence, find $f(s_n)$. Show that in each case $f(s_n)$ converges to $f(2)$.
   a) $s_n = 2$
   b) $s_n = 2 - 1/n$
   c) $s_n = \sqrt{4 + 1/n}$

THEORETICAL

7. Which of the sequences of Exercises 1 and 4 converge? Find the limit if the sequence converges. If the sequence does not converge, write *no limit*.

8. In each of the following, a specific function is given with a particular value of a. In each case find

$$\lim_{h\to 0} \frac{f(a+h)-f(a)}{h}.$$

a) $f(x) = x^2$, $a = 5$
b) $f(x) = x^2$, $a = 3$
c) $f(x) = x^2$, $a = -2$
d) $f(x) = x^2$, $a = -1/3$
e) $f(x) = x^2$, $a = b$
f) $f(x) = 3x$, $a = 7$
g) $f(x) = 4x$, $a = 6$
h) $f(x) = 17x$, $a = -3$
i) $f(x) = -3x$, $a = 12.4$

2.2 OPERATIONS WITH FUNCTIONS. CONTINUITY

We now introduce some commonly used notation to express more concisely some of the ideas presented in the previous section.

NOTATION: If a sequence s converges to some limit L, we denote this fact by

$$s_n \longrightarrow L.$$

If a function f has a limit L as x approaches a, we denote this fact by

$$\lim_{x\to a} f(x) = L.$$

It is human nature to want to deal primarily with subjects that are "predictable" and "well behaved." Since calculus deals with functions we might well ask which functions are predictable and well behaved. We examine this question in the following examples.

Example 11: Suppose that a function f is prescribed by the formula $f(x) = 2x$ for all numbers x other than 1; suppose we also know that $f(1)$ is defined and we want to know what it is. We would certainly conjecture that $f(1) = 2 \cdot 1 = 2$, since otherwise f would behave unpredictably when $x = 1$.

Looking at the graph of f (Figure 2–7), we see that the two parts of the graph of f for $x < 1$ and $x > 1$ can be joined to make a straight line by adding the point (1, 2) to the graph. Adding any point with x-coordinate 1 other than (1, 2) would leave the resulting graph with a break

A FUNCTION DEFINED BY $f(x) = 2x$ FOR ALL $x \neq 1$

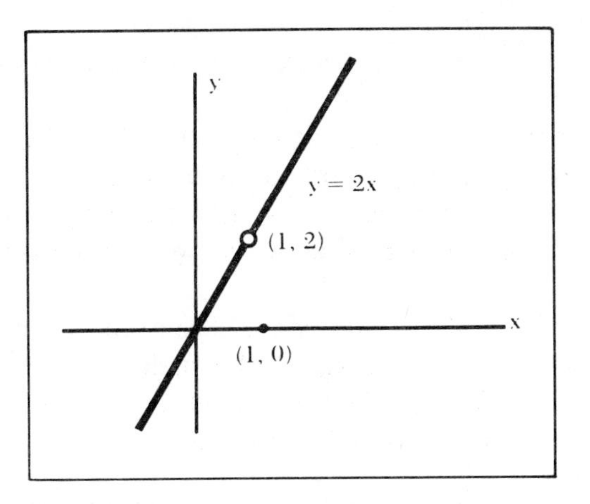

Figure 2–7 Only by defining $f(1) = 2$ can we connect the two parts of the graph of f.

where $x = 1$. Therefore, for the given function to behave in what seems to be the most reasonable way, the value at $x = 1$ should be $f(1) = 2$.

Example 12: Consider the function f defined by

$$f(x) = \begin{cases} -1, & \text{if} \quad x < 0, \\ 1, & \text{if} \quad x \geq 0. \end{cases}$$

The graph of f appears in Figure 2–8. We can reasonably say that f is not

A FUNCTION THAT IS NOT "PREDICTABLE" AT A POINT

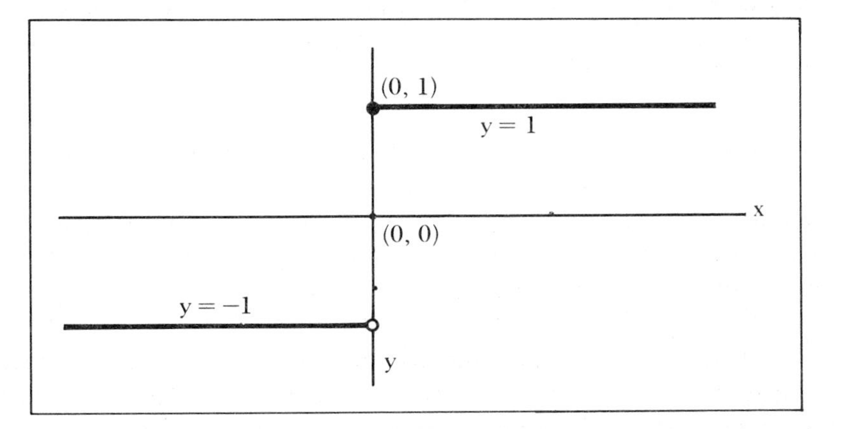

Figure 2–8 As x approaches 0 from the right, $f(x)$ approaches 1; as x approaches 0 from the left, $f(x)$ approaches -1. The values of $f(x)$ do not approach any one number as x approaches 0.

predictable as x approaches 0. If x approaches 0 from the positive direction, we expect $f(0)$ to be 1. However, if x approaches 0 from the negative direction, then the expected function value $f(0)$ becomes -1, *not* 1.

Note that there is no way to define $f(0)$ so that the graph of f forms a connected whole. It is important to understand, however, that there is nothing about f that contradicts its being a function; the point is that f is not as well behaved or as predictable as we might wish.

For "predictability" in the sense discussed in the two examples above, the graph of a function should be such that the two parts of the graph on either side of $(a, f(a))$ come together at $(a, f(a))$. This type of "joining" at $(a, f(a))$ is illustrated in Figure 2–9.

A CONTINUOUS FUNCTION IS "PREDICTABLE"

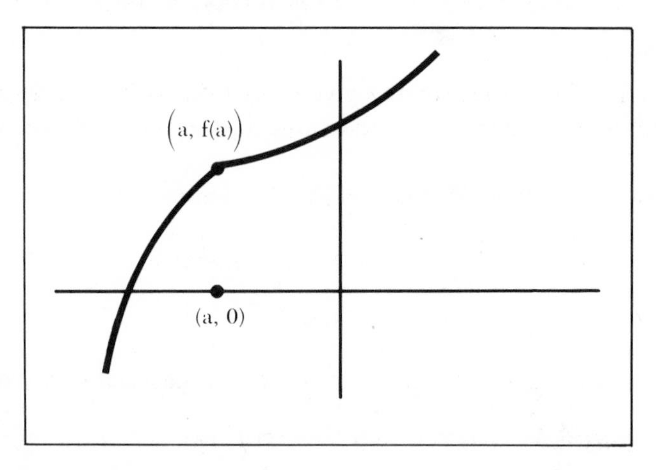

Figure 2–9 If f is continuous at $x = a$, then $f(x)$ approaches $f(a)$ as x approaches a. This appears graphically as the two parts of the graph of f on either side of $(a, f(a))$ joined by $(a, f(a))$.

If a graph is to be unbroken at $(a, f(a))$, the function values $f(x)$ must approach $f(a)$ as x approaches a. Note, in Figure 2–7, how the function value $f(x) = 2x$ approaches 2 as x passes through values approaching 1. In Figure 2–9 the values of $f(x)$ appear to approach $f(a)$ as x is chosen closer and closer to a. If $f(x)$ is to approach $f(a)$ as x approaches a, the following must hold:

i) $f(a)$ is defined, since there is no way that $f(x)$ can approach a nonexistent $f(a)$. Note that if f were not defined at a, there would then be a gap at $x = a$, and the two parts of the graph in Figure 2–9 would not be joined.

ii) $\lim_{x \to a} f(x)$ exists, since the function values are approaching some number (specifically $f(a)$) as x approaches a.

iii) $\lim_{x \to a} f(x)$ is $f(a)$. If $\lim_{x \to a} f(x) = L$ but L is not equal to $f(a)$, then we necessarily have a break in the graph at $x = a$ (Figure 2–10).

A FUNCTION WHICH IS NOT CONTINUOUS AT $x = a$

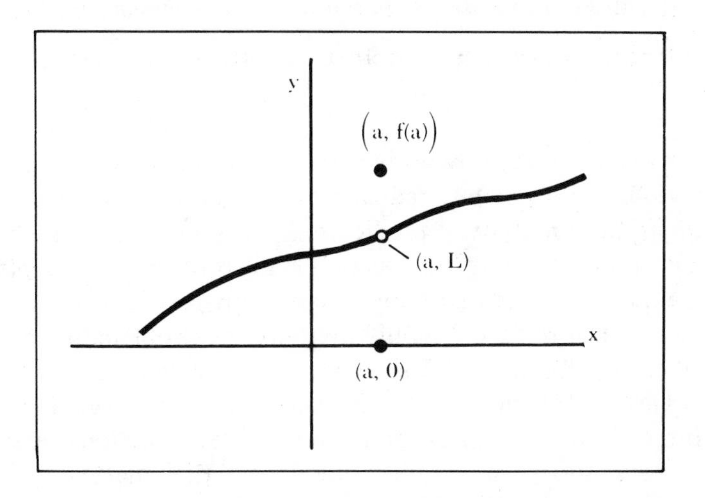

Figure 2–10 The function f is not continuous at $x = a$, because $f(x)$ approaches L as x approaches a but $L \neq f(a)$. This appears as a break in the graph of f.

We are now in a position to state what it means for a function to be well behaved in the sense implied in the previous discussion. Such well behaved functions are said to be *continuous.*

Definition 5: *A function f is said to be* ***continuous at x = a*** *if the above conditions (i) to (iii) apply to f at a. We say that f is* ***continuous*** *if it is continuous for each element in its domain.*

Example 13: The function $f(x) = 1/x$ is not continuous at 0, since $f(0)$ is not defined.

Example 14: If we define

$$f(x) = \begin{cases} x, & \text{if } x \neq 0, \\ 1, & \text{if } x = 0, \end{cases}$$

then f is not continuous at 0. Observe that $\lim_{x \to 0} f(x)$ exists and is equal to 0, but $0 \neq f(0) = 1$. Consequently, condition (iii) of Definition 5 is not satisfied and f is therefore not continuous. Sketch the graph of f to note the break that occurs at $x = 0$.

Example 15: The function f of Example 12 is not continuous at 0 since $\lim_{x \to 0} f(x)$ does not exist. What number we expect the limit to be depends on how we let x approach 0, and therefore f has no limit in Example 12.

We began this section by noting that it is human nature to want the functions we deal with to be predictable and well behaved, and we then went on to specify one type of predictability a function might have. It is also human nature to want to reduce difficult problems to simpler ones. By this time, graphing and exploring the basic properties of such functions as $f(x) = 3x - 1$ or $g(x) = 7$ should present little problem, but a function such as $m(x) = (x^3 - x + 1)^{25}$ is another matter, not to mention even more complicated functions we will encounter later. We therefore ask whether there is any way to examine complicated functions by reducing them to combinations of simpler functions. Fortunately, it is often possible to effect such a simplification, and, indeed, this constitutes one of the most important tools in applying the calculus.

Observe that $(x^3 - x + 1)^{25}$ is the quantity $x^3 - x + 1$ raised to the 25th power; $x^3 - x + 1$ is much simpler than $(x^3 - x + 1)^{25}$. Moreover, $x^3 - x + 1$ is a simple arithmetic combination, addition, of even simpler expressions, x^3, $-x$, and 1. We can carry the reduction still further: x^3 is the product of the quantity x with itself three times. The point is that $(x^3 - x + 1)^{25}$ consists of a number of arithmetic operations involving much simpler expressions. In the following definition we will see precisely how certain arithmetic operations with functions are performed.

Definition 6: *Let f and g be functions.*

i) The ***sum*** *$f + g$ of f and g is the function defined by $(f + g)(x) = f(x) + g(x)$, for all x for which both $f(x)$ and $g(x)$ are defined.*

ii) The ***product*** *fg of f and g is the function defined by $(fg)(x) = f(x)g(x)$, for all x for which both $f(x)$ and $g(x)$ are defined.*

iii) The ***quotient*** *f/g of f and g is the function defined by $(f/g)(x) = f(x)/g(x)$, for all x for which both $f(x)$ and $g(x)$ are defined and for which $g(x) \neq 0$.*

iv) Let r be a real number. The function rf is defined by $(rf)(x) = rf(x)$, for all x for which $f(x)$ is defined.

Example 16: The function $f(x) = x^{43} - 5x + 14$ is the sum of the functions x^{43}, $-5x$, and 14. (Strictly speaking, the expressions x^{43}, $-5x$, and 14 are not functions; however, it is clear what functions they represent.) We also see that x^{43} is the product of the function x with itself 43 times and that $-5x$ is the product of -5 and x.

Example 17: The function $f(x) = 1/x^2$ is the quotient of 1 divided by x^2. We might therefore gain information about f from information about 1 and x^2.

We would expect that if f and g are "well behaved," then the functions formed from f and g by simple arithmetic operations will also be well behaved. In paritcular, we note the following important result:

> If f and g are continuous at $x = a$, then $f + g$, fg, and rf for any real number r are also continuous when $x = a$; furthermore, f/g is continuous when $x = a$, provided $g(a) \neq 0$.

Thus, starting with continuous functions, we can produce other continuous functions by the arithmetic processes described in Definition 6 (Figure 2–11).

ADDITION OF FUNCTIONS

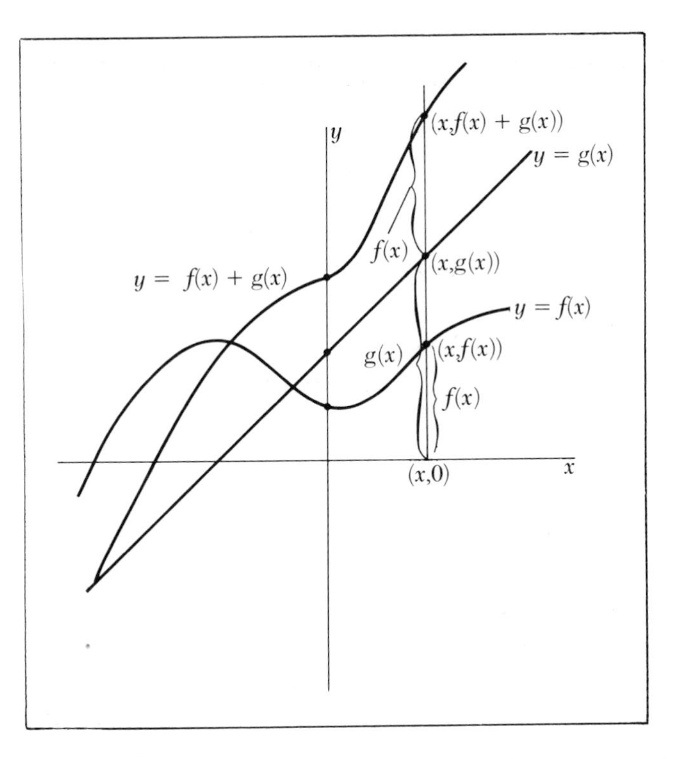

Figure 2–11 The graphs of f and g are "added" to get the graph of $f + g$. We obtain $(x, f(x) + g(x))$, which is on the graph of $(f + g)(x)$, by adding $(x, f(x))$ on the graph of f to $(x, g(x))$ on the graph of g.

There are also certain facts, which will be important to us later, about limits and arithmetic combinations of functions. Simply stated, these facts reduce to the principle that if f and g are functions that have limits L and L', respectively, as x approaches a, then the limit of the sum, product, or other arithmetic combination of f and g as x approaches a is found by performing the same arithmetic procedures with L and L'. Thus, for example,

$$\lim_{x\to a} (f+g)(x) = L + L';$$

the limit of the sum is the sum of the limits (Figure 2–12). More explicitly, these rules are as follows:

If $\lim_{x\to a} f(x)$ and $\lim_{x\to a} g(x)$ both exist, then:

i) $\lim_{x\to a} (f+g)(x) = \lim_{x\to a} f(x) + \lim_{x\to a} g(x)$,

ii) $\lim_{x\to a} (fg)(x) = \left(\lim_{x\to a} f(x)\right)\left(\lim_{x\to a} g(x)\right)$,

iii) $\lim_{x\to a} (rf)(x) = r \lim_{x\to a} f(x)$ for any real number r, and

iv) $\lim_{x\to a} (f/g)(x) = \left(\lim_{x\to a} f(x)\right)\Big/\left(\lim_{x\to a} g(x)\right)$, if $\lim_{x\to a} g(x) \neq 0$.

THE SUM OF THE LIMITS OF TWO FUNCTIONS IS THE LIMIT OF THEIR SUM

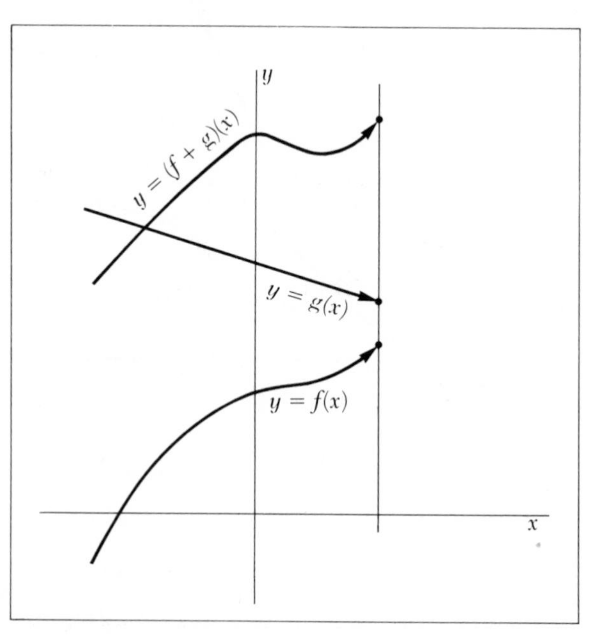

Figure 2–12 If $\lim_{x\to a} f(x)$ and $\lim_{x\to a} g(x)$ both exist, then $\lim_{x\to a} (f+g)(x)$ also exists and is equal to $\lim_{x\to a} f(x) + \lim_{x\to a} g(x)$.

Analogous statements hold about the limits of sequences. In order for these statements to be valid, it is *essential* that both $\lim_{x \to a} f(x)$ and $\lim_{x \to a} g(x)$ actually exist.

Example 18: Let $f(x) = (x^2 + 2x - 1)/x$. We will find $\lim_{x \to 1} f(x)$. Clearly, x itself has 1 as a limit when x approaches 1; x^2 has $1 \cdot 1 = 1$ as a limit; $2x$ has $2 \cdot 1 = 2$ as a limit; and -1, being a constant, has -1 as a limit as x approaches anything. Therefore, the limit of f as x approaches 1 is

$$(1 - 2 - 1)/1 = -2.$$

We also note that f, using arithmetic processes, is built up from continuous functions; hence f is continuous wherever it is defined—in this case, where $x \neq 0$.

Exercises

ROUTINE

1. For which sets of real numbers are each of the following functions continuous? Give an argument supporting each of your assertions; for example, show that the given function can be expressed as a simple arithmetic combination of continuous functions. For each point at which a function is discontinuous, state clearly why the function fails to be continuous at that point.

a) $f(x) = x$
b) $f(x) = 3x$
c) $g(x) = -9x^2 - 2$
d) $f(x) = 1/x$
e) $f(x) = 2x/(3 - x)$
f) $h(t) = (3t - 4)/(t^2 - 1)$
g) $f(x) = |x|$
h) $w(t) = t/(t^2 + 1)$
i) $g(x) = \begin{cases} x, & \text{if } x < 0, \\ 2x, & \text{if } x \geq 0 \end{cases}$

Warning: Do not be fooled by the fact that the definition of this function has two parts; sketch its graph.

j) $h(x) = \begin{cases} x, & \text{if } x \leq 0, \\ 2x + 1, & \text{if } x > 0 \end{cases}$

k) $m(t) = \begin{cases} t^2, & \text{if } t \neq 0, \\ 3, & \text{if } t = 0 \end{cases}$

Can this function be redefined at $x = 0$ so as to make it continuous?

2. Graph each of the following functions. Indicate clearly any points of the graph where discontinuities of the function occur.

a) $f(t) = 3t + 2$

b) $f(t) = -4t + 7$

c) $f(x) = \begin{cases} x^2, & \text{if } |x| \neq 1, \\ 1, & \text{if } |x| = 1 \end{cases}$

d) $f(x) = \begin{cases} x^2, & \text{if } |x| \neq 2, \\ 2, & \text{if } |x| = 2 \end{cases}$

e) $g(x) = x^3$

f) $h(x) = \begin{cases} x^2, & \text{if } x < 1, \\ x^3, & \text{if } x \geq 1 \end{cases}$

g) $(x) = \begin{cases} x^2, & \text{if } x < 4, \\ x^3, & \text{if } x \geq 4 \end{cases}$

CHALLENGING

3. Which of the following functions from "real life" situations are likely to be continuous?

a) the number of items a stamping machine has turned out in a time interval of length t

b) the weight of a human baby at time t after its birth

c) the amount of water in a reservoir t units of time after the reservoir has started filling

d) the speed of an airplane t units of time after take-off

e) the value of an acre of land located x miles from the center of the city of Chicago.

4. Let $f(x) = 3x + 1$ and $s_n = 1/n$.

a) Find $f(s_n)$.

b) To what number does the sequence of s_n converge? What is the limit of the sequence derived from $f(s_n)$?

c) What is the limit of $f(x)$ as x approaches 0? Why should we expect this limit to be the same as the limit of $f(s_n)$?

THEORETICAL

5. One can perform arithmetic operations with sequences just as with functions. In each of the following, two sequences, s and t, are defined.

i) In each case, find $s + t$ and st.

ii) In each case, find $s - 5t$ and s/t.

iii) Write out the first five terms of the sequences found in (i).

iv) Write out the first five terms of the sequences found in (ii).

v) The statements concerning limits of sequences are analogous to the statements concerning the limits of arithmetic combinations of functions. With this in mind, find the limits of each of the sequences found in (i).

vi) Find the limits of each of the sequences found in (ii).

a) $s_n = 1$ and $t_n = 2$

b) $s_n = 1/n$ and $t_n = 1 - 1/n$

c) $s_n = n$ and $t_n = (-1)^n n$

d) $s_n = 1/(n - \sqrt{2})$ and $t_n = n/(n^2 + 1)$

e) $s_n = 3n/(4n - 1)$ and $t_n = -6n/(7 - 5n)$

6. Employing the following procedure, one can determine where a continuous function has positive values and negative values.

i) If a function f

a) has function value of 0 at the end points of the closed interval $[a, b]$,

b) has no function values of 0 in the open interval (a, b), and

c) is continuous on $[a, b]$,

then:

if $f(x) > 0$ for *one* x in (a, b), then $f(x) > 0$ for *every* x in (a, b), and if $f(x) < 0$ for *one* x in (a, b), then $f(x) < 0$ for *every* x in (a, b).

ii) If a function f

a) is continuous for all real numbers $x < c$ or $c < x$, for some fixed real number c, and

b) if $f(x) \neq 0$ for all $x < c$ or $c < x$,

then:

if $f(x) > 0$ for one $x < c$ or $c < x$, then $f(x) > 0$ for all $x < c$ or $c < x$, or

if $f(x) < 0$ for one $x < c$ or $c < x$, then $f(x) < 0$ for all $x < c$ or $c < x$.

Using the information given in (i) and (ii), determine for which sets of real numbers the following functions have negative values and for which sets of real numbers they have positive values.

a) $f(x) = 2x$

b) $f(x) = 3x - 1$

c) $f(x) = (x - 2)(x - 3)$

d) $f(x) = (2x + 1)(3x - 2)$

e) $f(x) = (x - 3)(x - 5)(x + 7)$

f) $f(x) = x^2 + 1$

g) $f(x) = 1/(x - 1)$

7. Suppose that f is continuous at $x = a$; further, suppose that s is a sequence which converges to a, and for which $f(s_n)$ is defined for each term of s.

a) Since f is continuous, what number will $f(x)$ approach as values of x approach a?

b) Consider the sequence

$$f(s_1), f(s_2), f(s_3), \ldots$$

What number does s_n approach as n is chosen larger and larger?

c) Since s_n approaches a and the values $f(x)$ approach $f(a)$ as x approaches a, to what does the sequence in (b) converge?

Review of Chapter 2

A *sequence* is a list of numbers $s_1, s_2, s_3, \ldots,$ with one number s_n for each positive integer n. A sequence is said to *converge* to a *limit* L if all of the terms of the sequence approximate L with any desired accuracy once a specific term of the sequence is reached.

We say that a function f has a *limit* L *as* x *approaches* a if $f(x)$ approximates L with any desired accuracy, provided x is sufficiently close to a but $x \neq a$. A function is said to be *continuous at* a if $f(a)$ is defined and $\lim_{x \to a} f(x)$ exists and is equal to $f(a)$.

Functions and sequences can be added, multiplied, multiplied by a constant, or divided. The limit of any such arithmetic combination of sequences or functions is the corresponding arithmetic combination of the limits of the component functions, provided these limits exist and we are not dividing by zero (in the case of division). An arithmetic combination of continuous functions is continuous.

REVIEW EXERCISES

1. Find the limit, if a limit exists, of each of the following functions as x approaches 2.

a) $f(x) = x$
b) $f(x) = 3x$
c) $f(x) = -x - 1$
d) $f(x) = x^2 - 1$
e) $f(x) = 3$
f) $f(x) = 1/(x - 2)$
g) $f(x) = (x + 2)/(x - 2)$
h) $g(x) = x^2 + 3x + 0.6$
i) $h(x) = 1/(x^2 + 1)$
j) $f(x) = (x^2 - 4)/(x - 2)$
k) $f(x) = (1/x - 1/2)/(x - 2)$

2. Find the limit, if a limit exists, for each of the functions in Exercise 1 as:

a) x approaches -2,
b) x approaches 0,
c) x approaches a.

3. Which functions in Exercise 1 are continuous:

a) at $x = 0$?
b) at $x = 2$?
c) at $x = -2$?

4. Find the limit, if a limit exists, of each of the following sequences.

a) $s_n = n$

b) $s_n = \frac{1}{4}n$

c) $s_n = 1/(n + 3)$

d) $s_n = (-5n - 1)/(9n + 8)$

e) $t_n = (3n^2 + 1)/(-6n^2 + 7)$

f) $s_n = 3n/4n^2$

g) $s_n = (5n + 8)/(4n^3 - 0.9)$

h) $t_n = (5n^3 + 8)/(4n - 0.9)$

i) $s_n = (\sqrt{n} - 1)/(n - 1)$

Vocabulary

power function
exponent
exponential function
logarithm
logarithmic function

Chapter 3

Some Special Functions

3.1 THE POWER FUNCTION

We have already encountered functions of the type defined by

$$f(x) = x^n, \tag{1}$$

where n is a positive integer. To obtain $f(x)$ for a given value of x, we must multiply x by itself n times. Thus, for example,

$$2^7 = 2 \cdot 2 \cdot 2 \cdot 2 \cdot 2 \cdot 2 \cdot 2 = 128.$$

When n is a positive even integer, then x^n is always non-negative and the graph of $f(x) = x^n$ has the general form indicated in Figure 3–1. If n is a positive odd integer, then x^n is negative when x is negative and positive when x is positive. Hence, when n is odd, the graph of $f(x) = x^n$ has the general form indicated in Figure 3–2.

GRAPH OF $y = x^n$, n A POSITIVE EVEN INTEGER

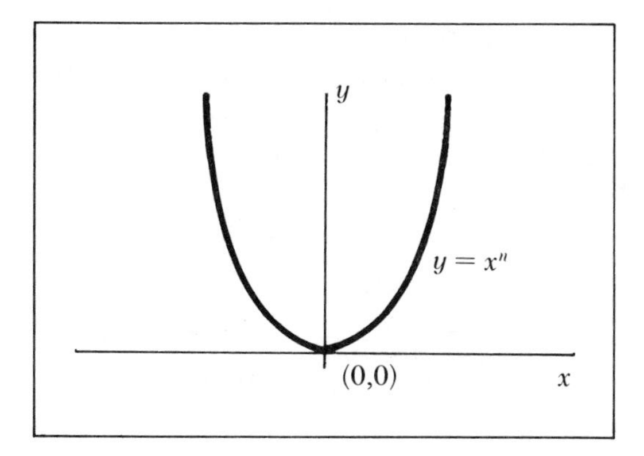

Figure 3–1 This graph is symmetric with respect to the y-axis and has its lowest point at (0, 0).

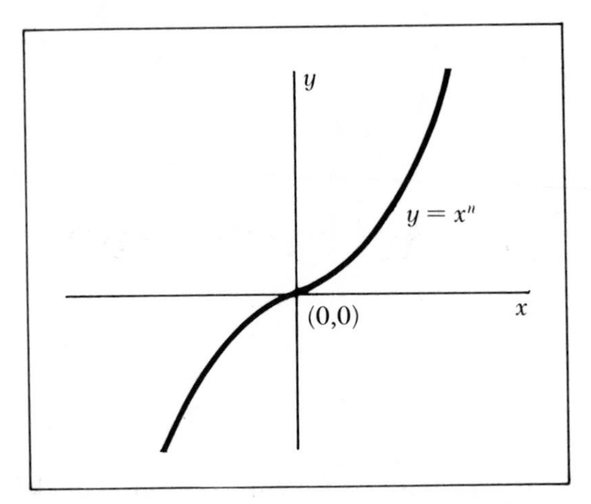

Figure 3–2 This graph is always rising; there is no highest or lowest point on the graph.

In this section we will define x^r, where r is some real number other than a positive integer. Our discussion will be based on the rules for dealing with exponents when the exponent is a positive integer. By assuming that these laws of exponents must hold even when the exponent is some real number other than a positive integer, we will be able to interpret such numbers as $5^{1/2}$ or $8^{-7/6}$.

The rules for exponents are as follows:

$$a^n a^m = a^{n+m}. \tag{2}$$

$$a^n b^n = (ab)^n. \tag{3}$$

$$(a^n)^m = a^{nm}. \tag{4}$$

These rules follow directly from the basic properties of multiplication of real numbers. To give an example, we now prove equation (2).

Example 1: Since

$$a^n = a \cdot a \cdot \ldots (n \text{ times})$$

and

$$a^m = a \cdot a \cdot \ldots (m \text{ times}),$$

$$a^n a^m = (a \cdot a \cdot \ldots (n \text{ times}))(a \cdot a \cdot \ldots (m \text{ times}))$$

$$= a \cdot a \cdot \ldots (m + n \text{ times}) = a^{n+m}.$$

Hence equation (2) is proved.

If we assume that equations (2), (3), and (4) should hold for powers of real numbers even when those powers are not positive integers, then we can interpret the meaning of non-integer powers.

Example 2: Consider $5^{1/2}$. According to equation (4),

$$(5^{1/2})^2 = 5^{1/2 \cdot 2} = 5^1 = 5.$$

Therefore $5^{1/2}$ is a number whose square is 5. Or, to express the result another way, $5^{1/2}$ is a square root of 5. By convention we take $5^{1/2}$ to be the positive square root of 5.

If a is any non-negative number, then

$$(a^{1/2})^2 = a^1 = a;$$

hence $a^{1/2}$ is a square root of a. We let $a^{1/2}$ be the non-negative square root of a.

We now generalize from Example 2. Suppose n is any positive integer. Then since

$$(a^{1/n})^n = a^{(1/n)n} = a^1 = a,$$

$a^{1/n}$ is an nth root of a. If a is positive and n is even, then a has two n^{th} roots, one positive and one negative. We will agree to let $a^{1/n}$ designate the positive n^{th} root. For example, $7^{1/8}$ designates the positive eighth root of 7.

If n is odd, then there is exactly one n^{th} root of a, and this root has the same sign as a. For example, $(-3)^{1/7}$ designates the seventh root of -3.

Since only non-negative real numbers have even roots, the function

$$f(x) = x^{1/n}$$

has the set of non-negative numbers as its domain if n is even; but since any real number has an nth root if n is odd, this same function has the entire set of real numbers as its domain if n is odd. The general forms of the graph of f are illustrated in Figures 3–3 and 3–4.

We now use equation (2) to obtain an interpretation for a zero power of a, where a is any non-zero real number. If (2) is to apply when n or m is zero, then

$$a^0 \cdot a^1 = a^{0+1} = a^1 = a.$$

Since $a \neq 0$, the only number by which we can multiply a and still obtain

GRAPH OF $y = x^{1/n}$, n A POSITIVE EVEN INTEGER

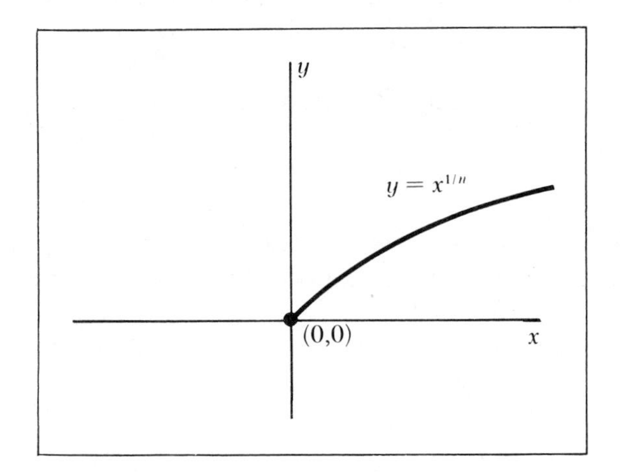

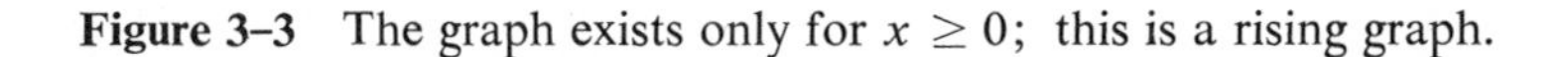

Figure 3–3 The graph exists only for $x \geq 0$; this is a rising graph.

GRAPH OF $y = x^{1/n}$, n A POSITIVE ODD INTEGER

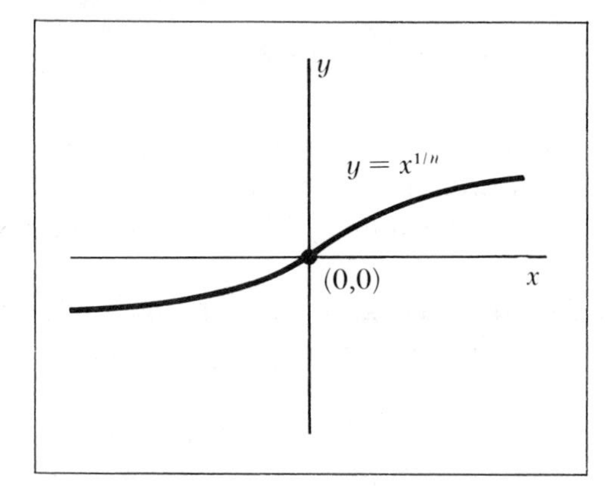

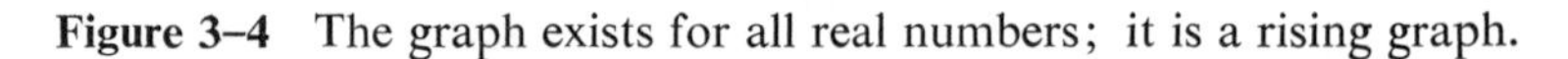

Figure 3–4 The graph exists for all real numbers; it is a rising graph.

a is 1; therefore it is reasonable to define

$$a^0 = 1$$

for any non-zero real number a.

We now investigate the meaning of a^{-t} where $a \neq 0$ and t is any number for which a^t is defined. In accordance with equation (2),

$$a^t a^{-t} = a^{(t-t)} = a^0 = 1.$$

It follows then that

$$a^{-t} = 1/a^t, \tag{5}$$

since $1/a^t$ is the only number by which we can multiply a^t and obtain 1.

Example 3: By the argument above,

$$5^{-1/2} = 1/5^{1/2}$$

and

$$7^{-3} = 1/7^3 = 1/343.$$

Example 4: According to rule (4),

$$(6^{-1})^{-1} = 6^{(-1)(-1)} = 6^1 = 6.$$

According to equation (5),

$$(6^{-1})^{-1} = 1/6^{-1} = 1/(1/6) = 6.$$

Thus we see that in this case equation (5) is in agreement with rule (4) for exponents.

Typical graphs of the function f defined by

$$f(x) = x^{-t}$$

appear in Figures 3–5 and 3–6.

GRAPH OF $y = x^{-1}$

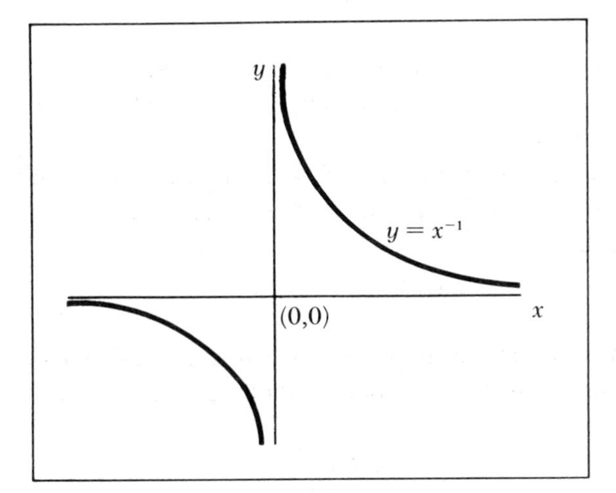

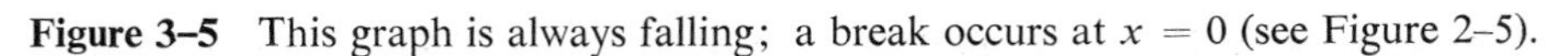

Figure 3–5 This graph is always falling; a break occurs at $x = 0$ (see Figure 2–5).

GRAPH OF $y = x^{-2}$

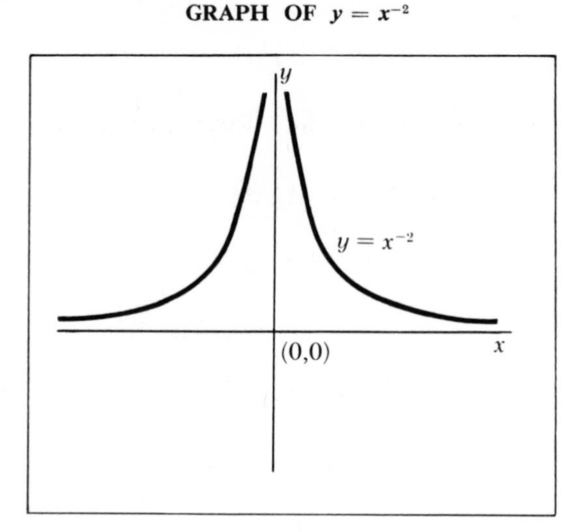

Figure 3–6 This graph is symmetric with respect to the y-axis; a break occurs at $x = 0$.

Finally, consider the number p/q, where p and q are positive integers. If a is any positive number, then by rule (4),

$$a^{p/q} = (a^{1/q})^p = (a^p)^{1/q}.$$

We can therefore interpret $a^{p/q}$ as either the qth root of the pth power of a, or as the pth power of the qth root of a.

Example 5: Let

$$f(x) = x^{2/3}.$$

Then $f(x)$ is the cube root of the square of x. Since every real number has a cube root, $f(x)$ is defined for every real number x. The graph of f appears in Figure 3–7.

If a is any positive number and r is an irrational number (one that cannot be expressed as the quotient of two integers), we may also define a^r as follows: Let

$$t_1, t_2, t_3, \ldots, t_n, \ldots$$

be the sequence of decimal numbers such that t_n represents r correct to n decimal places. Then the sequence t_n converges to r, and the sequence

$$a^{t_1}, a^{t_2}, a^{t_3}, \ldots$$

converges to a limit that we define to be a^r. In examples and applications we will, however, restrict our attention almost exclusively to exponents that are rational numbers.

GRAPH OF $y = x^{2/3}$

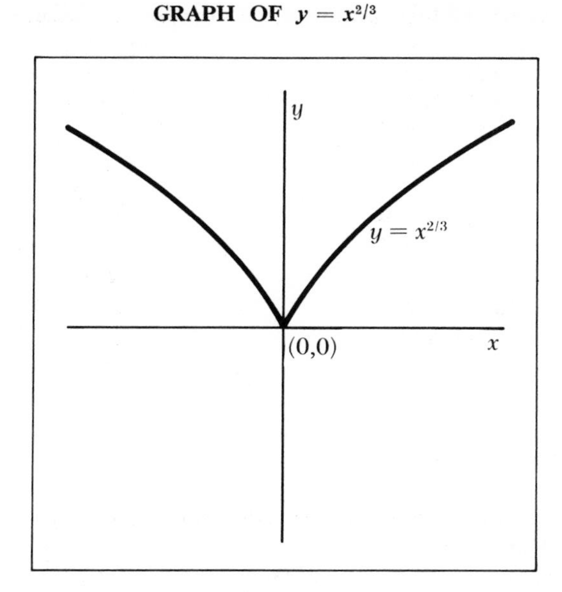

Figure 3–7 This graph is symmetric with respect to the y-axis; it appears to come to a point at its lowest point (0, 0).

Having discussed what we mean by a^t for real numbers a and t, we advance the following definition.

Definition 1: *The function f defined by*

$$f(x) = x^t,$$

where t is some real number, is called the ***power function with exponent t****, or simply a* ***power function.***

The power function is continuous for any x at which it is defined.

Exercises

ROUTINE

1. Evaluate each of the following:
   a) $9^{-1/2}$
   b) $16^{3/4}$
   c) 2^{-5}
   d) $1000^{-8/3}$
   e) $1^{0.09}$
   f) $(8^{0.2})^{-5/3}$
   g) $(9^{-1})^{-3}$
   h) $(0.01)^{9/2}$

2. Graph all functions in each set on the same coordinate axes:

a) $f(x) = x^0$	d) $f(x) = x^{-1}$
$f(x) = x^{1/4}$	$f(x) = x^{-2}$
$f(x) = x^{1/3}$	$f(x) = x^{-3}$
$f(x) = x^{1/2}$	$f(x) = x^{-4}$
b) $f(x) = x^0$	e) $g(t) = t^{1/4}$
$f(x) = x^{-1/4}$	$g(t) = t^{2/4}$
$f(x) = x^{-1/3}$	$g(t) = t^{3/4}$
$f(x) = x^{-1/2}$	$g(t) = t^{4/4}$
c) $f(x) = x$	f) $h(t) = t^{1/3}$
$f(x) = x^2$	$h(t) = t^{2/3}$
$f(x) = x^3$	$h(t) = t^{3/3}$
$f(x) = x^4$	$h(t) = t^{4/3}$

CHALLENGING

3. An art collector estimates that after n years a painting he just bought will be worth

$$1000 + 100n^{0.5} \text{ dollars.}$$

Sketch a graph that represents the value of the painting after n years. If the art collector's estimate is right, how much will the painting be worth after 25 years?

4. A manufacturer of sofas estimates that he will be able to sell

$$11{,}000 - 500t^{0.5}$$

units if each unit is priced at \$$t$.

a) Sketch a graph of $y = 11{,}000 - 500t^{0.5}$; y will be the number of units the manufacturer expects to sell if each sofa sells for \$$t$.

b) At what price per unit will the manufacturer find himself without any buyers?

THEORETICAL

5. Show that if a and b are any real numbers and n is a positive integer, then $a^n b^n = (ab)^n$.

6. Show that if a is any real number and n and m are positive integers, then $(a^n)^m = a^{nm}$.

3.2 THE EXPONENTIAL AND LOGARITHMIC FUNCTIONS

In various real life contexts one finds not only power functions but functions whose definition takes the form

$$f(x) = a^x, \tag{6}$$

where a is some positive real number.

Example 6: The *exponential density function*, which is of considerable importance in probability and statistics, has the form

$$f(x) = me^{-mx},$$

where m is a positive constant. We will have more to say about the number e later on. The point to be noted here is that f involves a variable power of a constant number e.

Definition 2: *Let a be a positive constant. A function whose definition takes the form*

$$f(x) = a^x$$

is called an ***exponential function*** *with* ***base*** *a.*

The general forms of the graphs of exponential functions appear in Figures 3–8, 3–9, and 3–10.

GRAPHS OF THE EXPONENTIAL FUNCTION $y = a^x$ FOR VARIOUS VALUES OF a

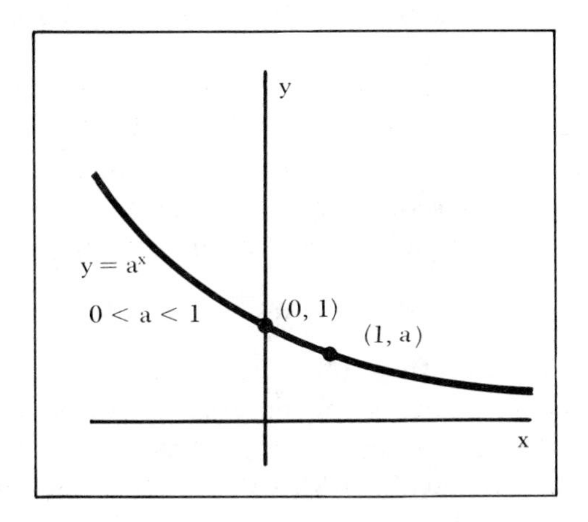

Figure 3–8 The graph of $y = a^x$ for $0 < a < 1$. This is a falling graph.

GRAPHS OF THE EXPONENTIAL FUNCTION $y = a^x$ FOR VARIOUS VALUES OF a (continued)

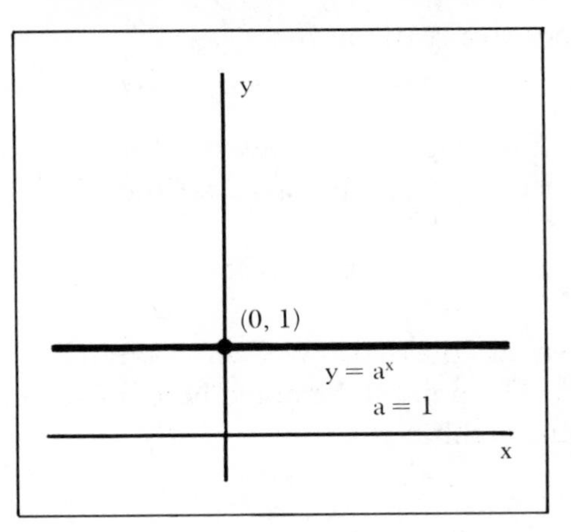

Figure 3–9 The graph of $y = 1^x$. This graph is a horizontal straight line, since $1^x = 1$ for any real number x.

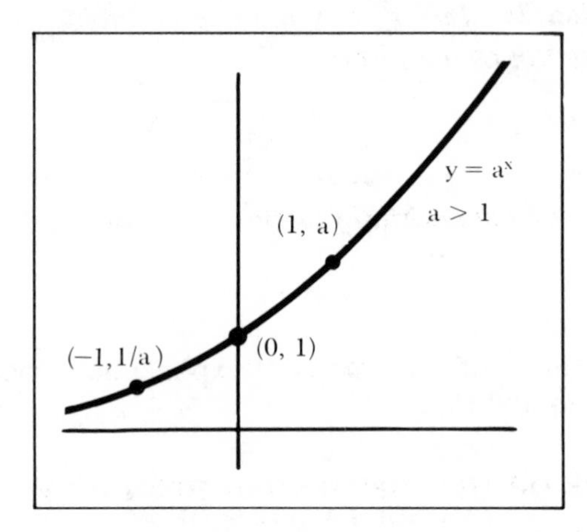

Figure 3–10 The graph of $y = a^x$ for $1 < a$. This is a rising graph.

Example 7: Certain figures associated with the economy or population, such as the total population or the Gross National Product, are often approximated by exponential functions, with time as the variable. For example, if the population of a given country at some initial time is 20,000,000, research might yield a population estimate of

$$P(t) = 20^{(1+0.01t)} \text{ million people,}$$

where t is the time in years from the initial time.

An exponential function is the natural function to describe many growth processes. In cell division, for example, new cells are produced and become available to divide still further. Thus, one cell divides into two, the two cells into four, the four into eight, and so forth. The number of cells increases exponentially—after n divisions there are 2^n cells—rather than linearly (in which case there would be $2n$ cells after n divisions).

An exponential function is continuous on the entire set of real numbers. It is strictly increasing if its base is greater than 1 and strictly decreasing if the base lies between 0 and 1 (cf. Figures 3–8, 3–9, and 3–10).

Closely related to the exponential functions with base a is the *logarithmic function with base a.*

Definition 3: *Let a be any positive real number other than 1. Then for any positive real number x, there exists precisely one number y such that*

$$x = a^y$$

(cf. Figures 3–8 and 3–10). We call y the ***logarithm to the base a of x****, and we denote this logarithm by*

$$\log_a x.$$

The function defined by

$$f(x) = \log_a x$$

is known as the ***logarithmic function*** *with* ***base a.***

By the definition of the logarithm to the base a, we can say that

$$x = a^{\log_a x}$$

and

$$\log_a (a^y) = y.$$

Note, however, that the logarithm to the base a of x is defined only if a is a positive number other than 1 and x is a positive number.

Example 8: Since $2^3 = 8$, $\log_2 8 = 3$.

Example 9: Since $3 = \sqrt{9} = 9^{1/2}$, $\log_9 3 = 1/2$.

Example 10: Since $a^0 = 1$ for any positive number a, $\log_a 1 = 0$—0 is the only power to which we can raise a and obtain 1—for any positive number a.

A logarithmic function is continuous on the set of positive real

GRAPHS OF $y = \log_a x$ FOR VARIOUS VALUES OF a

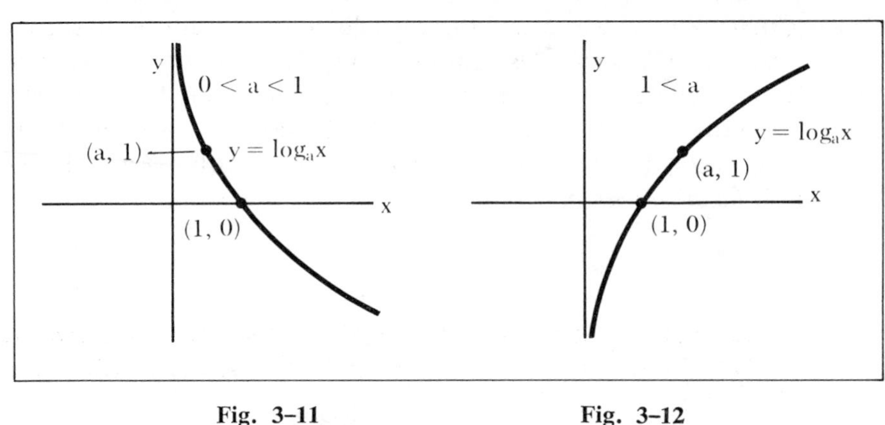

Fig. 3–11 Fig. 3–12

Figure 3–11 The graph of $y = \log_a x$ for $0 < a < 1$. Since $\log_a x$ is defined only for $x > 0$, the graph lies entirely to the right of the y-axis.
Figure 3–12 The graph of $\log_a x$ for $1 < a$.

numbers. The general forms of the graphs of the logarithmic functions are shown in Figures 3–11 and 3–12.

We now develop some of the basic properties of logarithms.

Example 11: By definition

$$x = a^{\log_a x} \quad \text{and} \quad y = a^{\log_a y}.$$

Therefore, by rule (2) for exponents,

$$xy = (a^{\log_a x})(a^{\log_a y}) = a^{\log_a x + \log_a y}.$$

But

$$xy = a^{\log_a xy}.$$

We therefore conclude that $\log_a xy$ and $\log_a x + \log_a y$ are both powers to which a can be raised to obtain xy. However, since there is only one power to which a can be raised to obtain xy, we conclude that

$$\log_a xy = \log_a x + \log_a y. \tag{7}$$

This means that the logarithm of a product is the sum of the logarithms.

Using the basic properties of exponents set forth in equations (2), (3), and (4), we can derive further fundamental properties of logarithms. One of these is the method for converting logarithms having one base into logarithms having another base.

Example 12: Let a and b be positive numbers other than 1. Then, since

$$a = b^{\log_b a} \quad \text{and} \quad x = b^{\log_b x} \quad \text{and} \quad x = a^{\log_a x},$$

we see that

$$b^{\log_b x} = x = a^{\log_a x} = (b^{\log_b a})^{\log_a x} = b^{(\log_b a)(\log_a x)}. \tag{8}$$

Comparing the first and last terms of equation (8) and recognizing that there is exactly one power to which b can be raised to obtain x, we observe that

$$\log_b x = (\log_b a)(\log_a x),$$

or

$$\log_a x = (1/\log_b a) \log_b x. \tag{9}$$

Thus, to obtain the logarithm to the base a of x from $\log_b x$, we merely divide $\log_b x$ by the constant $\log_b a$.

In Example 11 we saw that the use of logarithms can reduce multiplication to addition. Next we shall see that the use of logarithms can reduce the process of raising a number to a power to the process of multiplication.

Example 13: Since

$$x = a^{\log_a x},$$

it follows that

$$x^t = (a^{\log_a x})^t = a^{t \log_a x},$$

from which we conclude

$$\log_a (x^t) = t \log_a x.$$

Example 14: The number 10 is the base of the so-called *common logarithms*. (Logarithms to the base 10 are most suitable for computational purposes since our number system is based on 10.) Correct to 4 decimal places,

$$\log_{10} 2 = 0.3010$$

and

$$\log_{10} 3 = 0.4771.$$

From these two common logarithms we can compute the common logarithms of many other numbers. For example,

$$\log_{10} 6 = \log_{10} (2 \cdot 3) = \log_{10} 2 + \log_{10} 3 = 0.7781$$

(cf. Example 11). Also,

$$\log_{10} \sqrt{3} = \log_{10} 3^{1/2} = (1/2) \log_{10} 3 = 0.2385,$$

and

$$\log_{10}(3^7 2^{-6}) = \log_{10} 3^7 + \log_{10} 2^{-6} = 7\log_{10} 3 - 6\log_{10} 2 = 1.5337.$$

Since $\log_{10} 10 = 1$,

$$\log_{10} 2000 = \log_{10}(2 \cdot 10^3)$$
$$= \log_{10} 2 + 3\log_{10} 10 = \log_{10} 2 + 3 = 3.3010.$$

In general we find that

$$\log_{10}(a \cdot 10^n) = \log_{10} a + n,$$

where a is any positive real number.

Exercises

ROUTINE

1. Given that $\log_{10} 2 = 0.3010$, $\log_{10} 3 = 0.4771$, and $\log_{10} e = 0.4343$, compute each of the following.

a) $\log_{10} 16$
b) $\log_{10}(1/9)$
c) $\log_{10}(1/18)$
d) $\log_{10}(3/2)$
e) $\log_{10}(1/2)$
f) $\log_{10}(1/200)$
g) $\log_{10}(3600)$
h) $\log_2 3$
i) $\log_e 10$
j) $\log_e 6$
k) $\log_{10} 18^{1/7}$
l) $\log_{10} 6^{0.11}$
m) $\log_e (1/9)$
n) $\log_{10}(2/3)^{10}$
o) $\log_{10}(2^{0.4}/3^{0.9})$
p) $\log_{1/e} 10$
q) $\log_{10}(0.006)^{1/4}$
r) $\log_{0.1} 2^{19}$

2. Sketch a graph of each of the following functions and state precisely the domain of each function. Figures 3–8 through 3–12 indicate the general form of the graphs of logarithmic and exponential functions.

a) $f(x) = \log_{10} x$
b) $f(x) = \log_{0.1} x$
c) $f(x) = 10^x$
d) $g(x) = x + \log_{10} x$
e) $h(t) = 10^{-t}$ [Hint: $10^{-t} = (10^{-1})^t$.]
f) $f(x) = \log_2 x + \log_3 x$
g) $f(t) = \log_{10}(t^3)$
h) $f(x) = 10^x - 2^x$
i) $f(x) = \log_5 \sqrt{x}$
j) $f(x) = \log_{10}(1000x)$
k) $g(t) = \log_{10}(10^t)$

3. Sketch the graph of the exponential function that is given in Example 7.

THEORETICAL

4. Let a and b be positive numbers.

a) Show that $(a/b)^t = a^t/b^t = a^t b^{-t}$ for any real number t.

b) Show that

$$\log_{10} (a/b)^t = t(\log_{10} a - \log_{10} b).$$

5. An exponential relationship generally exists when some quantity is growing and the quantity added by growth also contributes to further growth. For example, people added to a population contribute to producing still more people. Find at least three real life examples in which an exponential function is likely to be operating.

3.3 FURTHER EXAMPLES AND APPLICATIONS

While logarithms to the base 10 are most useful for a purely computational viewpoint, logarithms to the base e, known as natural logarithms, find their application in real life situations that can be described by logarithmic functions. The exponential function with base e is also the most important of the exponential functions. The number e was introduced in Example 6 in connection with the exponential probability distribution; we will encounter it more frequently later on.

The value of e correct to 8 decimal places is

$$e = 2.71828183.$$

The number e is irrational; we can never represent it exactly in decimal form.

The number e is also the limit of the sequence defined by

$$s_n = (1 + 1/n)^n.$$

One application of this sequence is related to compound interest.

Example 15: One dollar is deposited in a fund that pays 100 per cent interest annually. If the interest is simply computed at the end of the year, then at the end of one year the depositor will have

$$\$1 + \$1(1.00) = \$2.$$

Suppose the interest is computed twice a year, using a 50 per cent rate each time. Then at the end of one year the account is worth

$$\$1 + \$1(0.50) + \$1.50(0.50) = \$(1 + 1/2)^2.$$

If the interest is computed quarterly, using a 25 per cent interest rate each time, it can be shown that the amount at year's end is

$$(1 + 1/4)^4.$$

In general, if the interest is compounded n times per year, at the end of the year the account will contain

$$\$(1 + 1/n)^n.$$

As n is chosen larger and larger, the amount in the account at year's end does not grow arbitrarily large, but rather approaches a limit. This limit is e. If the interest is compounded *continuously*, the account will contain e dollars at the end of the year.

More generally, it can be shown that if an account pays interest at rate r per year (where r is expressed as a fraction rather than as a percentage) and interest is compounded continuously, then \$1 at the beginning of the year will have grown to $\$e^r$ by year's end.

The number e also occurs frequently in probability and statistics.

Example 16: The *standard normal density function*, which is of central importance in probability and statistics, has the form

$$f(x) = (1/\sqrt{2\pi})e^{-x^2/2}.$$

We will now graph f.

Set

$$h(t) = (1/\sqrt{2\pi})e^t \quad \text{and} \quad g(x) = -x^2/2.$$

Then

$$f(x) = h(g(x)).$$

Since $h(t)$ is simply a constant multiplied by e^t, the graph of h has the form shown in Figure 3–13. The graph of $y = x^2/2$ appears in Figure 3–14; from this graph we conclude that the graph of g has the form shown in Figure 3–15.

We can determine the graph of f as follows: Given $x = a$, we compute $g(a)$. Then, letting $t = g(a)$, we calculate $h(t) = h(g(a))$. The point $(a, h(g(a))$ will be that point of the graph of f with first coordinate a. For example, if $x = 2$, then $g(2) = -2$ and $h(-2) = (1/\sqrt{2\pi})e^{-2}$; hence $(2, (1/\sqrt{2\pi})e^{-2})$ is the point of the graph of f with first coordinate 2. However, rather than working point by point, it is far more efficient to study the general properties of the graphs of g and h to obtain the general characteristics of the graph of f.

GRAPH OF $h(t) = (1/\sqrt{2\pi})e^t$

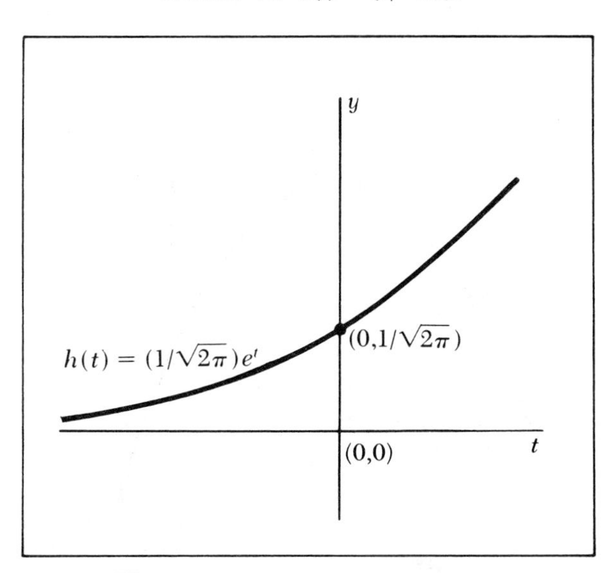

Figure 3–13 Since $(1/\sqrt{2\pi})e^t$ is simply a positive constant times e^t, the graph of h is essentially the same as the graph of $y = e^t$.

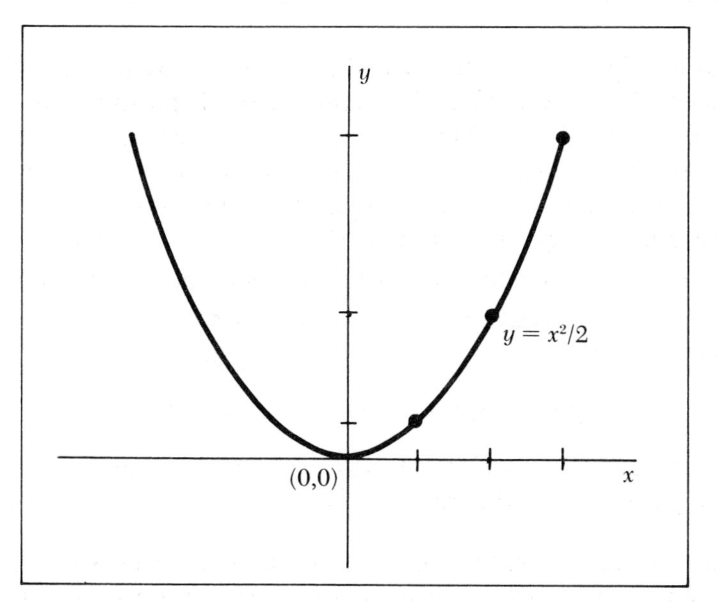

Figure 3–14 The graph of this function is a parabola above the x-axis.

GRAPH OF $y = -x^2/2$

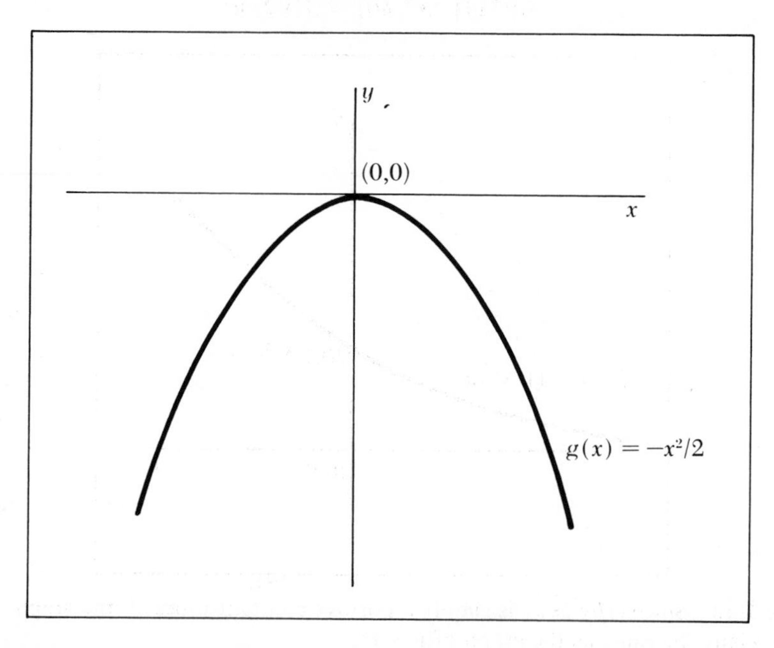

Figure 3–15 For each point $(x, x^2/2)$ on the graph of $y = x^2/2$, we have the corresponding point $(x, -x^2/2)$ on the graph of $y = -x^2/2$. This graph is the "reflection" of the one in Figure 3–14, with the x-axis as the "mirror."

For example, if x is positive and increasing, then $g(x)$ is negative and decreasing. If t is negative and decreasing, $h(t)$ is positive and decreasing toward 0 (Figure 3–16). Therefore, if x is positive and increasing, then $f(x) = h(g(x))$ is positive and decreasing toward 0. If x is negative and increasing toward 0, then $g(x)$ is negative and increasing toward 0; if t is negative and increasing toward 0, then $h(t)$ is positive and increasing toward $(1/\sqrt{2\pi})$. Therefore, if x is negative and increasing toward 0, then $f(x) = h(g(x))$ is positive and increasing toward $(1/\sqrt{2\pi})$. Finally, at $x = t = 0$, we see that

$$f(0) = h(g(0)) = (1/\sqrt{2\pi})e^0 = 1/\sqrt{2\pi}.$$

Note that $f(x)$ is positive for all real x and is maximum at $x = 0$. The graph of f is shown in Figure 3–17.

The natural logarithm function can be employed when the rate at which some quantity $Q(x)$ is changing is inversely proportional to x. We say that A is inversely proportional to B if

$$A = k/B,$$

where k is some constant; we illustrate such a case in the next example.

BEHAVIOR OF THE COMPOSITE FUNCTION $f(x) = (1/\sqrt{2\pi})e^{-x^2/2}$

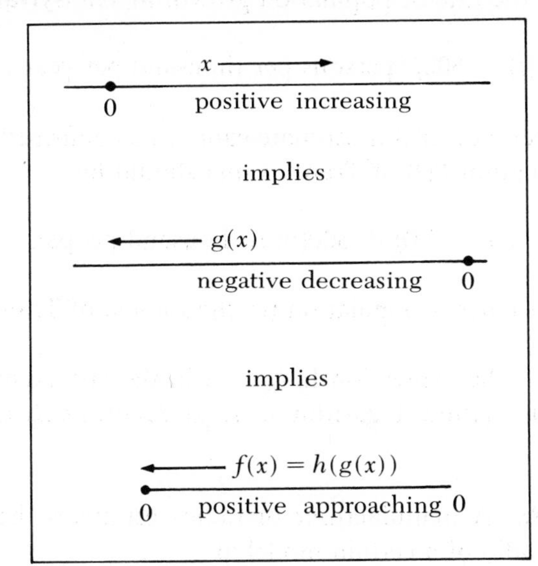

Figure 3–16 The chain of logical implications used to determine the behavior of the function f.

GRAPH OF $x = (1/\sqrt{2\pi})e^{-x^2/2}$

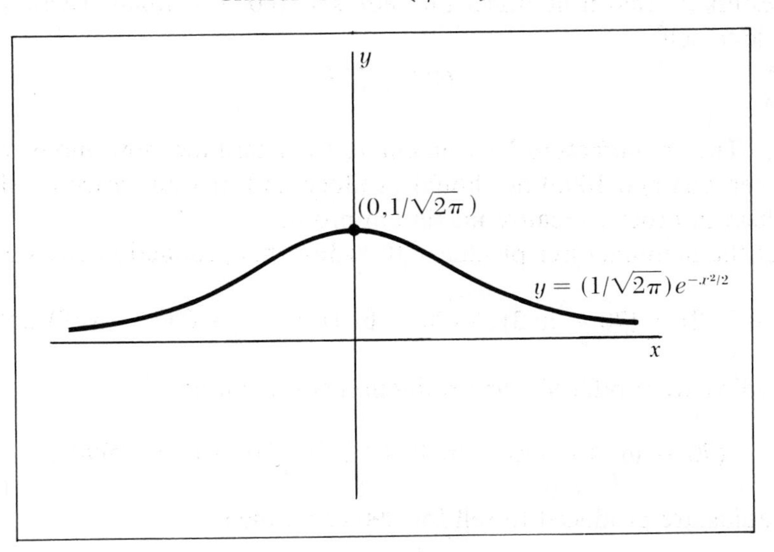

Figure 3–17 The information compiled about f is pictorially represented in this graph.

Example 17: The Census Bureau of Transylvania estimates that t years from now the rate of population growth in Transylvania will be

$$r(t) = 500/t \text{ persons per thousand per year.}$$

Although the accuracy of this estimate cannot be confirmed until later, in t years the population $P(t)$ of Transylvania should be

$$P(t) = P(0) + 500(\ln t) \text{ thousand people,}$$

where $P(0)$ is the current population (in thousands) of Transylvania.

NOTATION: The expression $\ln t$, used in the last example, is a short form of $\log_e t$, the natural logarithm of t. It denotes *only* a logarithm to the base e.

Example 18: A manufacturer of radios estimates the cost of producing the mth radio of a certain model at

$$70 - \ln(m + 2) \text{ dollars}$$

if any quantity from one to 1000 units are produced. The manufacturer also estimates that if he prices the radios at $100 - x$ dollars each, he will be able to sell

$$600 + 3^{0.1x}$$

units. The manufacturer's problem is to determine how many radios (between one and 1000) he should produce and at what price he should sell them in order to realize maximum profit.

If the manufacturer produces 10 radios, his production costs will be

$$(70 - \ln 2) + (70 - \ln 3) + (70 - \ln 4) + \ldots + (70 - \ln 10) \text{ dollars.}$$

If n radios are produced, the production costs will be

$$(70 - \ln 2) + (70 - \ln 3) + \ldots + (70 - \ln n) \text{ dollars.}$$

If n radios are produced to sell for \$90 each, then

$$600 + 3^{0.1(10)} = 600 + 3 = 603$$

radios will be sold, and revenues from the sales will total

$$603(90) = 54{,}270 \text{ dollars.}$$

The total profit will always be

$$\text{total sales revenue} - \text{production costs}$$

In subsequent chapters we will develop tools to solve the manufacturer's problem. In particular, we will learn how to determine how many radios should be produced to realize maximum profit if the radios are to sell for a certain price, as well as how to determine the price at which the radios should be sold to realize maximum profit if the manufacturer has decided to produce a certain number of them.

Exercises

ROUTINE

1. In each of the following groups, two functions f and g are defined. Using the same set of coordinates, graph f, g, and $y = f(g(x))$.

a) $f(t) = e^t$, $g(x) = x^2$
b) $f(t) = \ln t$, $g(x) = x^2$
c) $f(t) = e^t$, $g(x) = x^3$
d) $f(t) = \ln t$, $g(x) = x^{-3}$
e) $f(t) = e^t$, $g(x) = x + 1$
f) $f(t) = e^t$, $g(x) = 5x + 1$
g) $f(t) = \ln x$, $g(x) = 3x - 7$
h) $f(t) = t^4$, $g(x) = e^x$

CHALLENGING

2. Compute the profit of the radio manufacturer of Example 18 if he makes 10 radios and sells them for \$90 each.

3. What profit does the manufacturer of Example 18 make on the mth radio he produces if he sells each radio for \$80? What is the total profit he realizes on the first ten radios that he sells for \$80 each?

4. One hundred dollars is deposited in bank that pays 6 per cent interest per year compounded continuously. How much money is in the account at the end of one year?

THEORETICAL

5. Let

$$f(x) = xe^x.$$

a) For what values of x does $f(x) = 0$?
b) For what values of x is $f(x)$ positive?
c) For what values of x is $f(x)$ negative?

6. Let

$$f(x) = e^x \ln x.$$

a) For what values of x is $f(x)$ defined?
b) For what values of x is $f(x) = 0$?
c) For what values of x is $f(x)$ positive?
d) For what values of x is $f(x)$ negative?
e) Show that $f(x)$ increases as x increases.

Review of Chapter 3

The function f defined by

$$f(x) = x^t,$$

where t is some real number, is called the *power function* with *exponent t*. For any positive integer n, $x^n = x \cdot x \cdot \ldots$ (n times) and $x^{-n} = 1/x^n$. If x is any real number other than 0, then $x^0 = 1$. If m and n are non-zero integers, then $x^{m/n} = (x^{1/n})^m = (x^m)^{1/n}$; $x^{1/n}$ is an nth root of x.

A function whose definition takes the form

$$f(x) = a^x,$$

where $a > 0$, is called an exponential function with base a. The number e is one of the most important bases for exponential functions. If $a > 0$ and $x > 0$, then where $x = a^y$, the number y is called the *logarithm* of x to the base a. Graphs of the logarithmic function

$$f(x) = \log_a x$$

appear previously in our ditcussion.

REVIEW EXERCISES

1. Evaluate each of the following:

a) $4^{3/2}$
b) $4^{-3/2}$
c) $8^{1/3}$
d) $(4 \cdot 9)^{-1/2}$
e) $(3^2)^2$
f) $(7^{-2})^2$
g) $(5^{1/7})^7$
h) $(5^2 \cdot 16)^{-1/2}$

2. Given the following information, find each number asked for below. Note: All figures are approximate.

$\log_{10} 2 = 0.3$ $\qquad 10^{0.5} = 3.2$
$\log_{10} 3 = 0.48$ $\qquad e^{0.4} = 1.5$
$\log_e 10 = 2.3$ $\qquad e^{-1} = 0.37$

a) $\log_{10} 6$
b) $\log_{10} 4$
c) $\log_{10} 9$
d) $\log_{10} 81$
e) $e^{1.2}$
f) e^{-2}
g) $\log_{10} e$
h) $\log_{10} e^{0.4}$
i) $\log_{10} (3.2)$
j) $e^{-0.6}$
k) $\log_e 10^{-1}$
l) $\log_e 100$

3. Firms A and B both begin with sales of $10 million per year.
a) If A's and B's sales grow according to the following equations: $S_A(t) = 0.01t + 10$ and $S_B(t) = -0.01t + 10$, where t is measured in years and sales are in millions of dollars, which firm will have the greater sales after 3 years?
b) If the sales functions for A and B are $S_A(t) = e^{0.01t} + 10$ and $S_B(t) = \log_e (0.01t) + 10$, where t is again measured in years and sales in millions of dollars, which firm will have the greatest sales after 1 year, after 2 years, and after 3 years?

Vocabulary

derivative
tangent line
differentiable
composition
chain rule

Chapter 4

The Derivative

4.1 THE CONCEPT OF THE DERIVATIVE

In the last chapter we sought criteria for a "well behaved" function and formulated the notion of *continuity*. The following example shows that even some continuous functions are not as well behaved as we might wish.

Example 1: Consider the function $f(x) = |x|$, the graph of which appears in Figure 4–1. The function is continuous at $x = 0$, for as x approaches $0, f(x)$ also approaches 0. Suppose now that a car is traveling along the graph and its brakes and steering mechanism fail at the point (0, 0). How will the car leave the road? Running a finger along the graph should convince us that the answer to this question depends on the direc-

GRAPH OF $y = |x|$

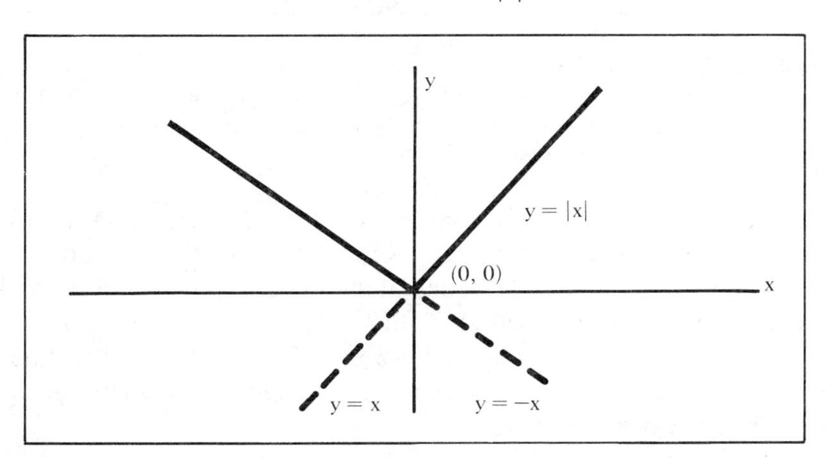

Figure 4–1 If a car is traveling along the graph toward (0, 0) and the constraints keeping the car on the graph fail at (0, 0), then the line along which the car leaves the graph will depend on the direction from which the car is approaching (0, 0).

tion from which the car is approaching (0, 0). If the car approaches from the right, then it will continue along the line with equation $y = x$; if the car approaches from the left, it will continue along the line with equation $y = -x$. Note that the graph has a sharp corner at (0, 0). A car would have trouble staying on the graph even if its brakes and steering mechanism did not fail.

Next we will examine a graph similar in shape to that of $f(x) = |x|$ but which remains "smooth" at (0, 0).

Example 2: Consider the function $g(x) = x^2$. The graph of g appears in Figure 4–2. Although the general shape of this graph is some-

GRAPH OF $g(x) = x^2$

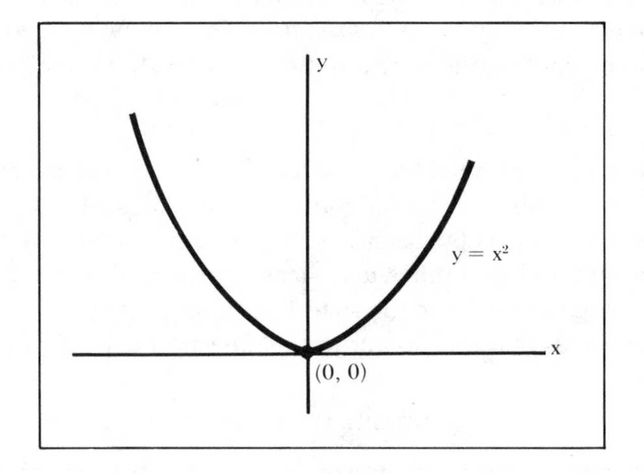

Figure 4–2 If a car is traveling along this graph toward (0, 0) and the constraints keeping the car on the graph fail at (0, 0), then the car will leave the graph along the x-axis regardless of the direction from which the car is approaching (0, 0).

what similar to the shape of the graph in Figure 4–1, this one has no sharp corner at (0, 0). We now seek an answer to the following question: What would happen if a car were traveling along the graph of g and its brakes and steering mechanism failed at (0, 0)? Along what line would the car leave the graph? In particular, we want to know whether the line on which the car would leave the graph is dependent on the direction from which the car approaches (0, 0).

According to Newton's Third Law of Motion, when the constraints that keep the car on the graph are removed, the car will proceed along the straight line *tangent* to the graph. We will therefore try to determine what line might reasonably be said to be tangent to the graph at (0, 0).

Consider a point (h, h^2), where $h \neq 0$, on the graph of $g(x) = x^2$ (Figure 4–3). The line that passes through (0, 0) and (h, h^2) approximates

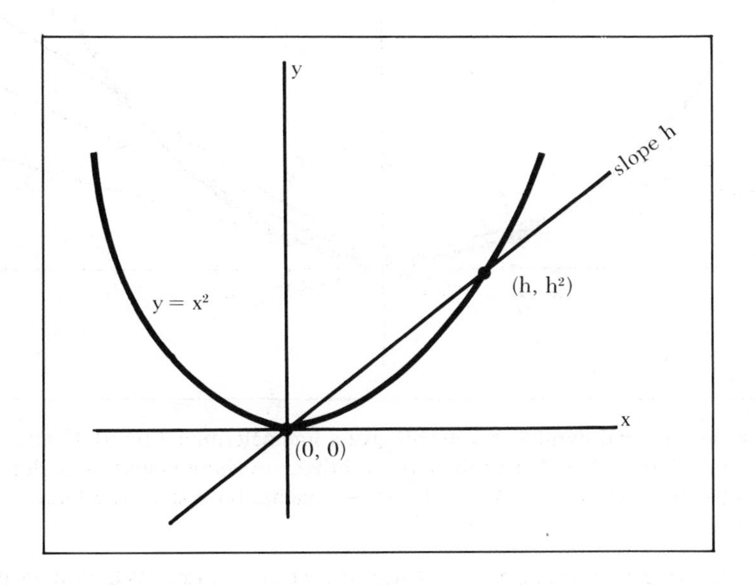

Figure 4–3 The line determined by (0, 0) and (h, h^2) has slope h. This line is an approximation to the line tangent to the graph at (0, 0).

the graph of $g(x) = x^2$ near (0, 0). The slope of this line is

$$\frac{h^2 - 0}{h - 0} = h^2/h = h, \tag{1}$$

regardless of whether h is positive or negative, just so long as $h \neq 0$. As h approaches 0, this slope approaches 0, and the line determined by (0, 0) and (h, h^2) looks more and more like a line that best approximates the graph of $g(x) = x^2$ at (0, 0) and the line we would expect the car to follow. It is also evident that equation (1) approaches 0 regardless of the direction from which h approaches 0. Therefore, the lines shown in Figure 4–4 are approaching the line with slope 0 which contains (0, 0), the x-axis, as h approaches 0 from either direction.

Thus, it seems reasonable to define the x-axis itself as the line tangent to the graph of $g(x) = x^2$ at (0, 0), and it is along this line that the car would travel if the restraint keeping it on the graph failed at (0, 0).

THE LIMITING POSITION OF A SEQUENCE OF LINES IS THE TANGENT LINE

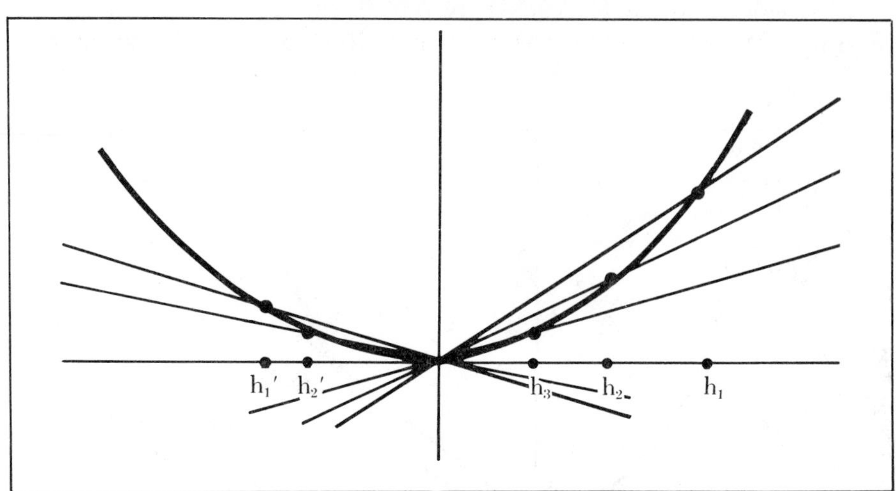

Figure 4–4 Any real number $h \neq 0$ will give a line determined by $(0, 0)$ and (h, h^2). As values of h, either positive or negative, are chosen closer and closer to 0, the line determined by $(0, 0)$ and (h, h^2) approaches the x-axis as a limit.

Example 3: Consider the function $f(x) = 1/x$. We will construct the line that best approximates the graph of f near the point $(1, 1)$. We use the same technique for the construction of the line that we used in Example 2. Select a point

$$(1 + h, 1/(1 + h)),$$

where $h \neq 0$, on the graph close to $(1, 1)$. We may choose h positive or negative, but we should choose $-1 < h$ so as to avoid the meaningless expression $1/0$. Then the line passing through $(1, 1)$ and $(1 + h, 1/(1 + h))$ approximates the graph of f at $(1, 1)$; such a line is pictured in Figure 4–5. The slope of this line is

$$\frac{1/(1 + h) - 1}{1 - (1 + h)} = \frac{1/(1 + h) - (1 + h)/(1 + h)}{h} = -1/(1 + h). \quad (2)$$

Now as h approaches 0, the limit of equation (2) is -1. Thus, as h approaches 0, the slopes of the approximating lines tend to -1 and the lines themselves tend toward the line that passes through $(1, 1)$ and has slope -1. The closer h is to 0, the better the approximation of the line determined by $(1, 1)$ and $(1 + h, 1/(1 + h))$ to the graph of f. The line with slope -1 passing through $(1, 1)$ is the best approximation. It is this line, therefore, that is most reasonably called the line *tangent* to the graph of f at $(1, 1)$.

APPROXIMATING THE GRAPH OF $f(x) = 1/x$ NEAR (1, 1)

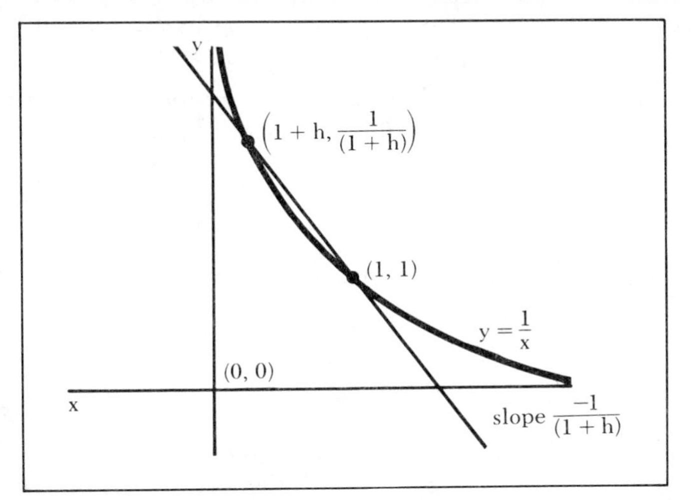

Figure 4–5 The line determined by (1, 1) and $(1 + h, 1/(1 + h))$ has slope $-1/(1 + h)$. This line approximates the graph of f near (1, 1).

Note that in both Examples 2 and 3 the slope of the tangent line was found as the limit of the slopes of other lines. This was necessary because two points are needed to determine a line. Merely specifying a point through which the tangent should pass does not suffice to determine it; indeed, we still have only a very informal notion of what a tangent line is. Merely estimating from a sketch of a graph what line best approximates the graph is too crude and qualitative to be of much practical value. Our method of finding specific lines related to the graph and the problem at hand, seeing how we can make such lines better and better approximations to a solution, and then accepting the limiting case of such approximating lines as the solution itself, gives us a well defined and precise answer in cases where an answer is possible.

Expressions similar to equations (1) and (2) arose in Example 10 of Chapter 2, as well as in the introductory discussion of Section 2.1. Recall that a ball was said to have traveled $s(t) = 16t^2$ feet t seconds after it had been dropped. We computed the average speed of the ball from time t_0 to time $t_0 + h$:

$$\frac{\text{total distance}}{\text{total time}} = 32t_0 + h \tag{3}$$

The limit of this expression as h approaches 0 is $32t_0$. Therefore, it is reasonable to speak of $32t_0$ as the *instantaneous speed* of the ball at time t_0. In this example we could confirm that $32t_0$ was in fact the speed of the ball at time t_0; the given data supported our tentative suggestion as to what the speed of the ball should be. In Example 10 of Chapter 2, a limit of the same form as equations (2) and (3) was considered in relation to the rate of change of function values; that example, and preferably all of Section 2.1, should at this point be reviewed.

From the examples in this section and Section 2.1, we see that the limit specified in the following definition appears in a number of contexts.

Definition 1: *Let f be a function such that*

$$\lim_{h \to 0} \frac{f(a+h) - f(a)}{h} \tag{4}$$

exists. Then we say that f is ***differentiable*** *at a, and we call expression (4) the* ***first derivative of f at a.*** *We generally denote the first derivative of f at a by $f'(a)$. The fraction $\dfrac{f(a+h) - f(a)}{h}$ in expression (4) is known as a* ***difference quotient.***

In order for the limit in (4) to exist, $f(a + h)$ must be defined for h

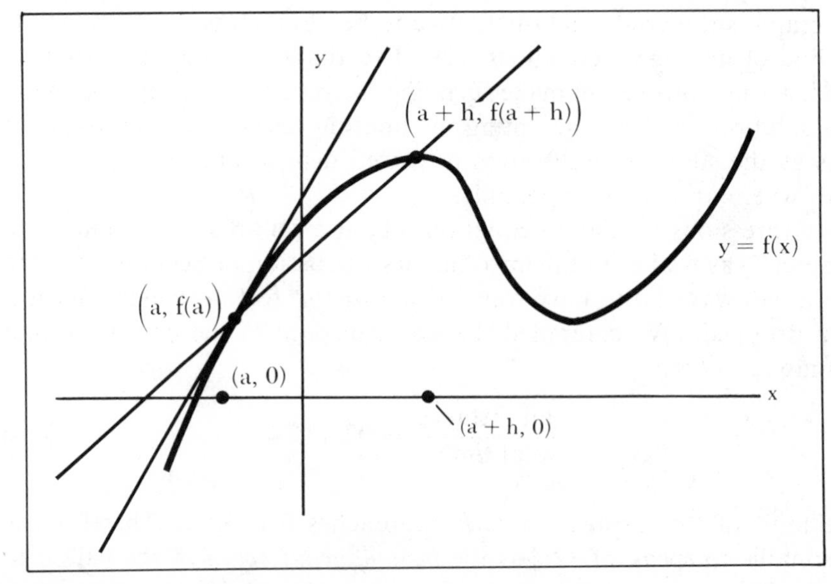

Figure 4–6 The tangent line to the graph of f at the point $(a, f(a))$ has slope $f'(a)$; it is the limit of the lines determined by the points $(a, f(a))$ and $(a + h, f(a + h))$ as h approaches 0.

in some open interval that contains 0; this in turn implies that f must be defined in some open interval containing a.

The word "first" in *first derivative* implies that there are second, third, and higher derivatives. Such is the case, as we will see later.

Considered geometrically, $f'(a)$ is the limit of the slopes of the lines passing through $(a, f(a))$ and $(a + h, f(a + h))$ as h approaches 0 (Figure 4–6). These lines will approach the line with slope $f'(a)$ passing through $(a, f(a))$ as h approaches 0. It is this latter line that we *define* to be the line tangent to the graph of f at $(a, f(a))$.

Definition 2: *Suppose that f is differentiable at a. We call $f'(a)$ the* ***slope*** *of f at a since $f'(a)$ is the slope of the line which best approximates the graph of f at $x = a$. We call the line with slope $f'(a)$ passing through $(a, f(a))$ the line* ***tangent*** *to the graph of f at $(a, f(a))$. (Thus the tangent line will have the equation*

$$y - f(a) = f'(a)(x - a).)$$

A function is said to be ***differentiable*** *if it is differentiable at each point of its domain.*

Note that we *define* the tangent line using $f'(a)$, even though we used the notion of a tangent line earlier to emphasize the importance of the derivative. The reason for this apparent circular reasoning is that we previously had only an informal and intuitive notion of what a tangent line is, based on previous experience either with tangents to circles in plane geometry or with the line a moving object will follow if no longer constrained to move in a certain path. Such intuitive notions are inadequate for obtaining a precise numerical characterization of the tangent line, the line of best approximation, or the rate of change. Nonetheless, we have used our intuitive notions to justify a particular mathematical computation.

The mathematical computation expressed in Definition 1 now allows us to develop more precise numerical formulations of the intuitive concepts that led to the definition of the derivative. If it turns out, however, that there are serious discrepancies between what we obtain from expression (4) and what our intuition leads us to expect, then we must conclude that either (4) is not adequate to define what our intuition proposes—and hence we should look for a new mathematical formulation of our intuitive ideas—or our intuition itself is in error. In Exercise 6 of Section 4.3 we will confirm that the tangent line defined in Definition 2 is the same tangent line encountered in the case of the circle in Euclidean plane geometry.

In the following example we test whether our formal definition of the derivative is in accord with our intuition.

Example 4: The graph of $f(x) = 3x - 7$ is a straight line with slope 3 (Figure 4–7). The line tangent to the graph at any point should be the line itself.

We can confirm this supposition intuitively by noting that if a car is traveling along a straight line and its steering mechanism fails, then unless other forces such as gravity or friction act to change the car's direction, it will continue to move along that same line.

For any real number a, in this case, we have

$$\frac{f(a+h) - f(a)}{h} = \frac{(3(a+h) - 7) - (3a - 7)}{h} = 3h/h = 3. \tag{5}$$

The limit of the difference quotient (5) as h approaches 0 is 3, because 3 is a constant and remains unchanged regardless of what h happens to be. The slope of f at $x = a$ is then 3, exactly the slope of the straight line that comprises the graph of f. Since the only straight line with slope 3 that passes through any point of the graph is the graph itself, we see that the graph is its own tangent line, as our intuition may have suggested.

Generally, we can show that if $f(x) = mx + b$, then the slope of f at a, which is equal to $f'(a)$, is m, the slope of the graph of f.

We conclude this section by demonstrating that the function in

GRAPH OF $f(x) = 3x - 7$

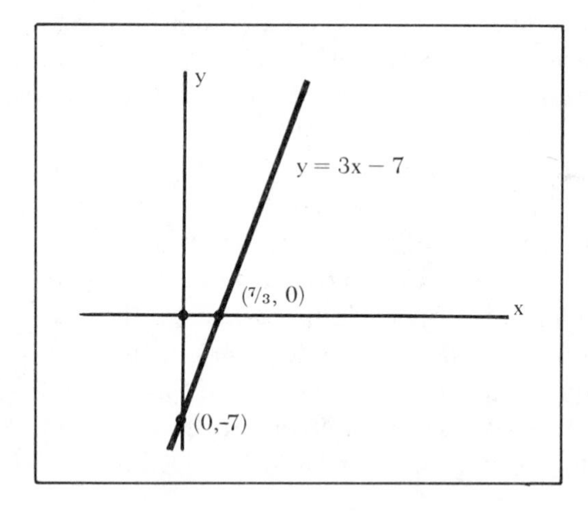

Figure 4–7 The graph of f is a straight line. In this case the line tangent to the graph at any point is the line itself.

Example 1 is not differentiable at 0; indeed, the "misbehavior" of this function is a manifestation of the nonexistence of $f'(0)$.

Example 5: In the case of $f(x) = |x|$, we find

$$\frac{f(0+h) - f(0)}{h} = \frac{|h|}{h}. \tag{6}$$

We now show that the difference quotient (6) has no limit as h approaches 0. As h approaches 0 from the positive direction, $|h|/h$ is 1 since $h > 0$ implies $|h| = h$. Therefore, we expect the limit of expression (6), if it exists, to be 1. However, as h approaches 0 from the negative direction, $|h| = -h$, and hence we would expect the limit of (6) to be -1. Since the limit of expression (6) as h approaches 0 depends on *how* h approaches 0, the limit does not exist. (The graph of $|h|/h$ is the same as that of the function in Example 12 of Chapter 2, but with the point (0, 1) deleted.) The nonexistence of this limit was geometrically evident: the line along which a car would travel off the graph at (0, 0) depends on the direction from which the car was coming.

In the next section we will develop simpler methods for calculating derivatives than by evaluating the limit of difference quotient (4) for each function we want to study.

Exercises

ROUTINE

1. Compute the first derivatives of the following functions at 0, at 2, and at -1. If a function is not differentiable at any of these points, write *not differentiable* and explain why it is not differentiable.

a) $f(x) = 2x$
b) $f(x) = 2x^2$
c) $f(x) = 2x^2 - 1$
d) $f(x) = 2x^2 + 8$
e) $f(x) = -7x + 7$
f) $f(x) = x^{-1}$
g) $f(x) = x^{1/2}$
h) $f(x) = x^{-1/2}$
i) $f(x) = 2x^3$
j) $f(x) = -2x^3$
k) $h(x) = x^2 + 2x + 3$
l) $g(x) = x^2 - 5x + 9$
m) $f(x) = -7x^2 - 5x + 9$
n) $f(x) = x/(x + 1)$
o) $f(x) = |x| + x$
p) $f(x) = |x| - x$

2. Find the tangent line to the graph of each of the functions listed in Exercise 1 at the point where $x = 0$. Sketch the graphs of each of the functions together with the graph of the tangent line at $x = 0$.

CHALLENGING

3. A rocket is moving straight upwards in such a way that it is exactly $80t^2$ miles from its launch point t minutes after launching.
   a) According to results discussed earlier, particularly at the beginning of Section 2.1, the speed at time t_0 will be the first derivative of $f(t) = 80t^2$ at time t_0. Find a function with variable t to express the speed of the rocket at time t. [Hint: First show that $f'(a) = 160a$ for $t = a$; $f'(a)$ is the speed of the rocket at time a. What function gives the speed of the rocket at time t?]
   b) Find the speed at which the rocket is moving 1 minute after launching, 5 minutes after launching, and 17.5 minutes after launching.

THEORETICAL

4. We have already seen that continuous functions are fairly "well behaved." Differentiable functions are even better behaved; their graphs are not only "connected," but also "smooth," and we have powerful techniques for studying their properties. The following steps are intended to show that a differentiable function is also continuous.
   a) Explain why a function f is continuous at a if and only if $\lim_{h \to 0} f(a + h) = f(a)$. Recall that the definition for continuity states that f is continuous at a if $\lim_{x \to a} f(x) = f(a)$.
   b) Examine the difference quotient in expression (4). Show that the numerator of this difference quotient will approach 0 as h approaches 0 if and only if f is continuous at a. Therefore, if the limit of the difference quotient as h approaches 0 is to exist, f must be continuous at a.

5. Show that expression (4) is the same as $\lim_{x \to a} \dfrac{f(x) - f(a)}{x - a}$. [Hint: Let $x = a + h$.]

6. a) Determine m and b such that $f(x) = mx + b$ best approximates $g(x) = x^{1/2}$ near $x = 1$.
   b) Use the function you obtain in part (a) to approximate $(1.01)^{1/2}$. Find the actual value of $(1.01)^{1/2}$ to check the accuracy of your approximation.

4.2 METHODS FOR COMPUTING DERIVATIVES

Naturally we do not want to go through the lengthy process of computing the limit of the difference quotient (4) every time a derivative is to be found, particularly for a complicated function. Fortunately, there are rules that often greatly simplify the process of finding the first derivative. We will omit detailed proofs for most of the rules we state, but we will try to make them reasonable when this can be done without too much formality.

First, instead of computing the first derivative of a function f separately at each of several points, it is generally possible to obtain a function, which we denote by f', whose value at x is the first derivative of f at x if f is differentiable at x.

Example 6: If $f(x) = 3x - 7$, then, as we saw in Example 4, $f'(a) = 3$ for any real number a. Therefore, f' in this case is the function defined by $f'(x) = 3$. For any real number x, $f'(x)$ is in fact exactly what its notation implies: it is the first derivative of f at x.

Example 7: Suppose $f(x) = x^2$. Then, evaluating expression (4) at x to obtain the first derivative of f at x, we see that

$$\lim_{h \to 0} \frac{f(x+h) - f(x)}{h} = \lim_{h \to 0} \frac{(x+h)^2 - x^2}{h} = \lim_{h \to 0} 2x + h = 2x.$$

Therefore, $f'(x) = 2x$.

The function f' referred to above is generally called the *first derivative* of f. Instead of computing the first derivative of f at individual points, we almost always try to find f', the first derivative function.

If the first derivative of a function is to measure the rate of change of a function, then a function that is constant, and hence whose value does not change, should have 0 as its first derivative at every point. In the following proposition we see that such is the case and that the logical consequences of the formal definition of the first derivative are in accord with our intuitive understanding of its meaning.

Proposition 1: If $f(x) = k$ and k is a constant, then $f'(x) = 0$.

PROOF:

$$f'(x) = \lim_{h \to 0} \frac{f(x+h) - f(x)}{h} = \lim_{h \to 0} \frac{k - k}{h} = \lim_{h \to 0} 0/h = 0.$$

Therefore, $f'(x) = 0$ for every real number x.

In Section 2.1 we defined certain arithmetic operations involving functions with the idea that these operations would facilitate the solution of certain kinds of problems later on. In the next several propositions we examine how the derivatives of certain arithmetic combinations of differentiable functions are related to the derivatives of their various component functions.

Proposition 2: If r is any real number and f is a function that is differentiable at a, then $(rf)'(a)$ exists and is equal to $rf'(a)$; therefore, $(rf)'(x) = rf'(x)$. In other words, the first derivative of a constant multiplied by a function f equals the constant multiplied by the first derivative f' of f.

PROOF:

$$\frac{(rf)(a+h) - (rf)(a)}{h} = \frac{rf(a+h) - rf(a)}{h} = r\left(\frac{f(a+h) - f(a)}{h}\right). \tag{7}$$

Now the limit of $(f(a+h) - f(a))/h$ as h approaches 0 is $f'(a)$; therefore the limit of the last expression in equation (7) as h approaches 0 is $rf'(a)$. This is the same as the limit of the first expression in (7) as h approaches 0, which is, by definition, $(rf)'(a)$. Therefore, $(rf)'(a) = rf'(a)$.

Example 8: We saw in Example 7 that if $f(x) = x^2$, then $f'(x) = 2x$. Therefore, if $g(x) = 5x^2$, then $g'(x) = 5(2x) = 10x$.

Proposition 3: If f and g are both differentiable at a, then $f + g$ is also differentiable at a and $(f+g)'(a) = f'(a) + g'(a)$: the derivative of a sum is the sum of the derivatives of the components.

PROOF:

$$\begin{aligned}
(f+g)'(a) &= \lim_{h\to 0} \frac{(f+g)(a+h) - (f+g)(a)}{h} \\
&= \lim_{h\to 0} \frac{(f(a+h) + g(a+h)) - (f(a) + g(a))}{h} \\
&= \lim_{h\to 0} \left[\frac{(f(a+h) - f(a))}{h} + \frac{(g(a+h) - g(a))}{h}\right] \\
&= \lim_{h\to 0} \frac{f(a+h) - f(a)}{h} + \lim_{h\to 0} \frac{g(a+h) - g(a)}{h}.
\end{aligned}$$

The last line is justified by the fact that the limit of a sum is the sum of the limits of the components when the limits exist. Finally, by the definition of the first derivative,

$$(f+g)'(a) = f'(a) + g'(a). \tag{8}$$

Example 9: Let $h(x) = x + x^2$. Since $h(x)$ is the sum of x and x^2, $h'(x)$ is the sum of the derivatives of x and x^2; these derivatives are 1 and $2x$, respectively. Therefore, $h'(x) = 1 + 2x$.

So far, the results concerning operations with derivatives have taken similar forms: If f and g are both differentiable at a and we are to perform some arithmetic operation on f and g, then that combination of f and g has as its derivative the same arithmetic combination of the derivatives of f and g. Thus, the derivative of the product of a constant and a function is the product of the constant and the derivative of the function, and the derivative of a sum is the sum of the derivatives. Unfortunately, this general rule, which was valid in the case of limits, breaks down when applied to the products and quotients of functions. The derivative of a product is *not* the product of the derivatives; neither is the derivative of a quotient the quotient of the derivatives. We illustrate this point in the next example.

Example 10: Let $f(x) = x$ and $g(x) = 2x$. Then $(fg)(x) = 2x^2$, so $(fg)'(x) = 4x$. On the other hand, $f'(x)g'(x) = 2$ since $f'(x) = 1$ and $g'(x) = 2$. Thus, in this case we observe that the derivative of the product is *not* the product of the derivatives. The following proposition provides the correct expression for the derivative of the product of two functions.

Proposition 4: If f and g are both differentiable at a, then fg is also differentiable at a and

$$(fg)'(a) = f(a)g'(a) + f'(a)g(a):$$

the first derivative of fg is f times the first derivative of g plus g times the first derivative of f.

PROOF:

$$\begin{aligned}(fg)'(a) &= \lim_{h \to 0} \frac{(fg)(a+h) - (fg)(a)}{h} \\ &= \lim_{h \to 0} \frac{f(a+h)g(a+h) - f(a)g(a)}{h}.\end{aligned} \tag{9}$$

From the form of equation (9) it is evident that there is no reason to expect that $(fg)'(a)$ should be $f'(a)g'(a)$. By a series of algebraic manipulations, which we will omit, equation (9) can be expressed as

$$(fg)'(a) = \lim_{h\to 0}\left[f(a+h)\,\frac{g(a+h)-g(a)}{h} + g(a)\,\frac{f(a+h)-f(a)}{h}\right]. \tag{10}$$

We may confirm the accuracy of this assertion by actually expanding and adding the fractions in equation (10). The limit expressed in equation (10) is, by definition, $f(a)g'(a) + g(a)f'(a)$, which proves the proposition.

Example 11: Let $f(x) = x^4$. Since $f(x)$ is the product $x^2 \cdot x^2$, and we have already established that the first derivative of x^2 is $2x$, we find, using Proposition 4, that $f'(x) = x^2(2x) + (2x)x^2 = 2x^3 + 2x^3 = 4x^3$.

Example 12: Let $f(x) = (x^2 + 7x - 1)(x + 3)$. Using Propositions 1, 2, and 3, we can easily verify that the first derivatives of $x^2 + 7x - 1$ and $x + 3$ are $2x + 7$ and 1, respectively. Therefore, using Proposition 4, we find that $f'(x) = (x^2 + 7x - 1)(1) + (x + 3)(2x + 7)$.

The rule for finding the derivative of a quotient of two functions is stated in the following proposition.

Proposition 5: Suppose that the functions f and g are both differentiable at a and that $g(a) \neq 0$. Then f/g is differentiable at a, and

$$(f/g)'(a) = \frac{g(a)f'(a) - g'(a)f(a)}{(g(a))^2}. \tag{11}$$

We make no attempt now to justify Proposition 5; we will outline its proof after introducing the Chain Rule in the next section.

Example 13: Consider $f(x) = 1/x$. Since $f(x)$ is the quotient of 1 divided by x, and since 1 and x have the derivatives 0 and 1, Proposition 5 indicates that $f'(x) = (x \cdot 0 - 1 \cdot 1)/x^2 = -1/x^2$.

Example 14: Consider $f(x) = x/(x + 1)$. Clearly, $f(x)$ is the quotient of x divided by $x + 1$, and the derivatives of x and $x + 1$ are both 1. Therefore, by Proposition 5,

$$f'(x) = ((x + 1)(1) - (1)(x))/(x + 1)^2 = 1/(x + 1)^2.$$

We will lack rules for differentiating most powers of x. For example, we have no easy means of differentiating $f(x) = x^{89.1}$. Neither have we rules for differentiating the exponential and logarithmic functions; indeed, we are not yet certain that these functions are differentiable. We now address ourselves to the question of differentiating these functions. First we consider powers of x.

Proposition 6: Suppose $f(x) = x^r$, where r is a fixed real number. Then $f'(x) = rx^{r-1}$.

We make no attempt to justify Proposition 6 generally. Note, however, that we have already confirmed Proposition 6 in certain special instances; for example, we found the first derivative of x^2 to be $2x$, and the first derivative of x^4 to be $4x^3$ (Example 11), and the first derivative of x^{-1} to be $(-1)x^{-2}$ (Example 13).

Example 15: Let $f(x) = x^{1/2}$, $x > 0$. Then $f'(x) = (1/2)x^{-1/2} = 1/(2\sqrt{x})$. Note that even though $f(x)$ is defined for $x = 0$, $f'(0)$ is not defined. Geometrically, $f(x) = x^{1/2}$ fails to be differentiable at $x = 0$ because the tangent line to the graph of f at (0, 0) is the y-axis (see Figure 4–8), the slope of which is undefined.

GRAPH OF $y = x^{1/2}$

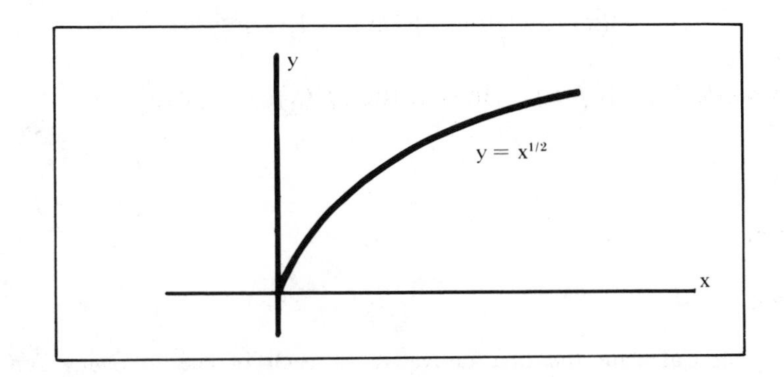

Figure 4–8 This graph has the y-axis as a tangent line at (0, 0); the slope of the y-axis is undefined. The function $f(x) = x^{1/2}$ fails to be differentiable at (0, 0).

The exponential and logarithmic functions with base e have particularly advantageous properties with respect to differentiation.

Proposition 7: Let $f(x) = e^x$ and $g(x) = \log_e x$ (recall that $\log_e x$ is defined only for $x > 0$). Then:

a) $f'(x) = e^x$ for every real number x: the first derivative of e^x is itself; and
b) $g'(x) = 1/x$ for every positive real number x.

PROOF: The proof of assertion (a) is complicated for the purposes of this discussion. We will provide a partial proof of assertion (a) in the exercise section that follows. We can provide justification—though not a proof—of assertion (b), given that (a) is true, once the Chain Rule has been introduced.

The Chain Rule will also enable us to find the derivatives of exponential functions with bases other than ė. We conclude this section by showing how to obtain the derivatives of logarithmic functions with bases other than e.

Example 16: Suppose that $f(x) = \log_a x$. Then from equation (12) of Section 2.3 we find

$$f(x) = (1/\log_e a) \log_e x.$$

Thus we see that $f(x)$ is simply a constant, $1/\log_e a$, multiplied by $\log_e x$. Since the first derivative of $\log_e x$ is $1/x$, we apply Proposition 2 and obtain

$$f'(x) = (1/\log_e a)(1/x) = 1/(x \log_e a).$$

Example 17: If $f(x) = \log_5 x$, then $f'(x) = 1/(x \log_e 5)$.

Exercises

ROUTINE

1. Calculate the first derivative of each of the following functions. Indicate any points at which the function itself is defined but the first derivative is not; for example, $x = 0$ is such a point for the function $f(x) = x^{1/2}$ (cf. Example 15).

a) $f(x) = -5x$
b) $g(x) = 7x + 1$
c) $f(x) = 3x^2$
d) $g(x) = x^{1/2}$
e) $f(t) = 1/t$
f) $f(x) = x^3$
g) $f(x) = x^2 - 3x + 2$
h) $w(t) = t^2 + 5t - 1$
i) $f(x) = x^2 - 1/x$
j) $g(x) = x/(x + 1)$
k) $f(x) = 1/(x^2 + 1)$
l) $g(t) = t/(t^2 - 1)$
m) $g(x) = x^3 - x^2$
n) $f(x) = x^3 - x^2 + 1$
o) $f(x) = x^{0.9}$
p) $f(x) = e^x$
q) $g(x) = \log_e x$
r) $f(x) = (1/x) + \log_e x$
s) $f(t) = 1/(t \log_3 t)$
t) $h(x) = x^2 + 3x + e^x$
u) $g(t) = 3t^2 + 5 + e^t \log_e t$
v) $f(t) = 5t^5e^t + 4t^4 \log_e t$

2. Try to determine where the first derivative of each function listed in Exercise 1 has a positive value, where it has a negative value, and where it is 0.

CHALLENGING

3. For each function listed in Exercise 1, try to sketch the graph of the function together with the graph of its first derivative on the same set of axes; this may, however, be difficult in certain instances, given the limited techniques of graphing introduced so far.

4. Employing the laws of exponents, Propositions 4 and 5, and anything else in this section that you find useful, compute the first derivative of each of the following functions.

a) $f(x) = e^{2x}$
b) $f(x) = e^{-x}$
c) $g(x) = e^{\log_e x}$
d) $f(t) = e^{-3t+1}$
e) $g(t) = e^2$
f) $f(x) = \log_3 e^x$

THEORETICAL

5. Suppose that $g(x) = (f(x))^2$, where f is a differentiable function. Use Proposition 4 to show that $g'(x) = 2f(x)f'(x)$; then use this result and Proposition 4 to obtain the first derivative of $h(x) = (f(x))^3$. What do you think is the first derivative of $g(x) = (f(x))^n$, where n is any positive integer?

6. Consider $f(x) = e^x$. To find $f'(x)$, we must calculate

$$\lim_{h\to 0} \frac{e^{x+h} - e^x}{h} = \lim_{h\to 0} e^x \left(\frac{e^h - 1}{h}\right)$$

$$= e^x \lim_{h\to 0} \left(\frac{e^h - 1}{h}\right).$$

Derive each expression in this series of equalities from the one preceding it, showing why the equalities are in fact valid. It can be shown, although we will not do so here, that

$$\lim_{h\to 0} \left(\frac{e^h - 1}{h}\right) = 1.$$

Try to verify experimentally that this limit is reasonable. For example, compute $(e^h - 1)/h$ for h equal to 0.1, 0.01, 0.001, and 0.0001; observe how the values obtained approach 1 as h decreases.

4.3 COMPOSITION OF FUNCTIONS. THE CHAIN RULE

Thus far we have used only arithmetic operations to build more complex functions from simpler ones. We now introduce another function operation, *composition*, which will substantially simplify the process of finding the derivatives of many types of functions.

A function g associates a real number $g(x)$ with each number x in its domain. If $g(x)$ is in the domain of another function f, then we can find $f(g(x))$, the function value of f at the number $g(x)$. This process is illustrated in Figure 4–9; we also illustrate it now using several examples.

THE COMPOSITION OF THE FUNCTION f WITH THE FUNCTION g

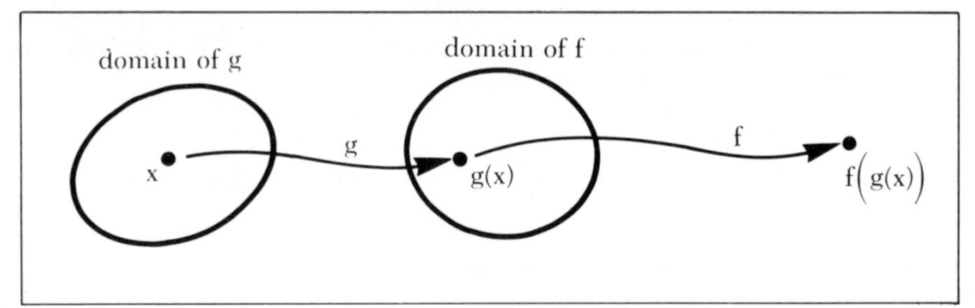

Figure 4–9 The function g associates $g(x)$ with x; the function f associates $f(g(x))$ with $g(x)$. Thus, the composition of f with g associates $f(g(x))$ with each element x in the domain of g for which $g(x)$ is in the domain of f.

Example 18: Suppose $f(x) = x^2$ and $g(x) = x - 7$. Then g is the rule which tells us to subtract 7 from each real number in its domain. The function f squares each number in its domain. Therefore,

$$f(g(x)) = f(x - 7) = (x - 7)^2,$$

and, similarly,

$$g(f(x)) = g(x^2) = x^2 - 7.$$

Example 19: Let $f(x) = e^x$ and $g(x) = x^2$. Then $f(g(x)) = f(x^2) = e^{x^2}$, while $g(f(x)) = g(e^x) = (e^x)^2 = e^{2x}$.

Example 20: Let $h(t) = (t^3 + 34t^2 - 1)^{45}$. We can express $h(t)$ as $f(g(t))$, where $g(t) = t^3 + 34t^2 - 1$ and $f(t) = t^{45}$. Note that $g(t)$ provides the number $t^3 + 34t^2 - 1$; f then raises this number to the 45th power. We have expressed the rather cumbersome function h in terms of the much simpler functions f and g.

Definition 3: *Let f and g be functions. Then the function h defined by $h(x) = f(g(x))$ is called the* ***composition of f with g.*** *We sometimes denote the composition of f with g by $f \circ g$. Thus,*

$$(f \circ g)(x) = f(g(x)).$$

Note that $(f \circ g)(x)$ is defined only for those real numbers x such that $g(x)$ is in the domain of f. Also note that $f \circ g$ and $g \circ f$ are usually quite different, as is illustrated in Examples 18 and 19.

Example 21: Suppose $f(x) = x^{1/2}$ and $g(x) = x - 2$. Then

$$(f \circ g)(x) = f(g(x)) f(x - 2) = (x - 2)^{1/2}.$$

Although $g(1)$ is defined and equals -1, $f(g(1))$ is not defined, since $f(-1) = (-1)^{1/2}$ is not defined.

In Example 19 we observed that the function $s(x) = e^{x^2} = f(g(x))$, where $g(x) = x^2$ and $f(x) = e^x$. The previous section contains no rule enabling us to compute $s'(x)$, yet s itself is the composition of two functions whose first derivatives are easily found. We therefore ask whether it is possible to express the first derivative of s in terms of f and g and their first derivatives. To answer this question, we adopt the following procedure.

Rather than restrict f and g to the specific functions in the paragraph above, we let f and g be general differentiable functions, and we set

$$s(x) = f(g(x)).$$

We want to obtain an expression for $s'(x)$ in terms of $f(x)$, $f'(x)$, $g'(x)$, and $g'(x)$. First, we know that

$$s'(x) = \lim_{h \to 0} \frac{f(g(x + h)) - f(g(x))}{h}. \qquad (12)$$

We now rewrite the fractional part of the right side of equation (12):

$$\frac{f(g(x+h))-f(g(x))}{g(x+h)-g(x)}\cdot\frac{g(x+h)-g(x)}{h}. \tag{13}$$

Next we change the notation in order to see what happens to the first term of equation (13) as h approaches 0. Clearly, the second term of equation (13) has $g'(x)$ as its limit as h approaches 0.

Set $g(x) = a$ and $g(x + h) = y$. Since g' exists, g is continuous; hence $g(x + h)$ approaches $g(x)$ as h approaches 0. Therefore we see that

$$\lim_{h\to 0}\frac{f(g(x+h))-f(g(x))}{g(x+h)-g(x)} = \lim_{y\to a}\frac{f(y)-f(a)}{y-a}. \tag{14}$$

The right side of equation (14), however, is simply $f'(y)$ (cf. Exercise 4 of Section 2.3). When we substitute $g(x)$ for y, equation (14) becomes $f'(g(x))$. Therefore the limit of equation (13) as h approaches 0 is

$$f'(g(x))g'(x),$$

and it is this expression that equals $s'(x)$. We summarize our conclusion:

Proposition 8: If $s(x) = f(g(x))$, and if f and g are differentiable functions, then

$$s'(x) = f'(g(x))g'(x). \tag{15}$$

Proposition 8 is known as the *Chain Rule;* it is an indispensable tool for calculating derivatives.

Example 22: If $s(x) = e^{x^2}$, then $s(x) = f(g(x))$, where $g(x) = x^2$ and $f(x) = e^x$. Since $f'(x) = e^x$, we see that $f'(g(x)) = f'(x^2) = e^{x^2}$. Also, $g'(x) = 2x$. Therefore, by the Chain Rule, $s'(x) = f'(g(x))g'(x) = e^{x^2}2x = 2xe^{x^2}$.

Example 23: We saw in Example 20 that $h(t) = (t^3 + 34t^2 - 1)^{45} = f(g(t))$, where $g(t) = t^3 + 34t^2 - 1$ and $f(t) = t^{45}$. Now $g'(t) = 3t^2 + 68t$ and $f'(t) = 45t^{44}$. Therefore, according to the Chain Rule, $h'(t) = f'(g(t))g'(t) = f'(t^3 + 34t^2 - 1)(3t^2 + 68t) = 45(t^3 + 34t^2 - 1)^{44}(3t^2 + 68t)$.

Example 24: Suppose $g(x) = e^{f(x)}$, where f is a differentiable function. Then since $g(x)$ is the composition of e^x with $f(x)$, we see by the

Chain Rule that $g'(x) = e^{f(x)}f'(x)$. We now use this fact to prove Proposition 7(b), which was stated on page 110.

Let $f(x) = \log_e x$. By definition,

$$x = e^{\log_e x} = e^{f(x)}. \tag{16}$$

The first derivative of x is 1, while the first derivative of $e^{f(x)}$, as we have seen, is $e^{f(x)}f'(x)$. Since both sides of equation (16) are equal, their first derivatives are also equal. Hence

$$1 = e^{f(x)}f'(x) = e^{\log_e x}f'(x) = xf'(x). \tag{17}$$

Solving equation (17) for $f'(x)$, we see that $f'(x)$, the first derivative of $\log_e x$, is $1/x$.

Example 25: Suppose $p(x) = (g(x))^r$, where r is some fixed real number and g is a differentiable function. Then $p(x)$ is the composition of $g(x)$ with x^r. Since the first derivative of x^r is rx^{r-1}, then by the Chain Rule $p'(x) = r(g(x))^{r-1}g'(x)$. In particular, if $p(x) = 1/g(x) = (g(x))^{-1}$, then $p'(x) = (-1)(g(x))^{-2}g'(x)$. Using this fact, we can now prove Proposition 5, the rule for finding the first derivative of the quotient of two functions.

Let $h(x) = f(x)/g(x)$; then $h(x) = f(x)(g(x))^{-1}$. We have expressed h as the *product* of two functions. Therefore, applying Proposition 4 we obtain

$$\begin{aligned} h'(x) &= f(x)(-1)(g(x))^{-2}g'(x) + f'(x)(g(x))^{-1} \\ &= \frac{f'(x)}{g(x)} - \frac{f(x)g'(x)}{(g(x))^2} \\ &= \frac{f'(x)g(x) - f(x)g'(x)}{(g(x))^2}, \end{aligned}$$

which is the expression given for $h'(x)$ in Proposition 5.

We already know how to differentiate $f(x) = e^x$; we conclude this section by showing, with the help of the Chain Rule, how to differentiate exponential functions with bases other than e.

Example 26: Consider $f(x) = a^x$, where a is a positive real number. Then $a = e^{\log_e a}$. Therefore,

$$f(x) = a^x = (e^{\log_e a})^x = e^{x \log_e a}.$$

We note that $f(x)$ has the form $e^{g(x)}$, where $g(x) = x \log_e a$. Therefore, by the Chain Rule, $f'(x) = e^{g(x)}g'(x) = e^{x\log_e a} (\log_e a) = (e^{\log_e a})^x(\log_e a) = a^x \log_e a$. Note that since $x \log_e a$ is simply the constant $\log_e a$ multiplied by x, its derivative is just $\log_e a$. We conclude that if $f(x) = a^x$, then $f'(x) = a^x \log_e a$.

Most of this chapter has been devoted to methods of finding the first derivative. In the next chapter we will learn more about what information the first derivative gives about a function, as well as about a number of practical applications of the derivative.

Table 4.1 lists some basic types of functions and their derivatives for ready reference.

TABLE 4.1 *FIRST DERIVATIVES OF COMMON FUNCTIONS*

	Function	***First Derivative***
1.	$f(x) = k$, k a constant	$f'(x) = 0$
2.	$f(x) = x^r$, r a constant	$f'(x) = rx^{r-1}$
3.	$f(x) = kg(x)$, k constant	$f'(x) = kg'(x)$
4.	$f(x) = g(x) + h(x)$	$f'(x) = g'(x) + h'(x)$
5.	$f(x) = g(x)h(x)$	$f'(x) = g(x)h'(x) + h(x)g'(x)$
6.	$f(x) = g(x)/h(x)$	$f'(x) = (h(x)g'(x) - g(x)h'(x))/(h(x))^2$
7.	$f(x) = g(h(x))$	$f'(x) = g'(h(x))h'(x)$
8.	$f(x) = \log_e x$	$f'(x) = 1/x$
9.	$f(x) = e^x$	$f'(x) = e^x$
10.	$f(x) = a^x$, $a > 0$	$f'(x) = a^x \log_e a$
11.	$f(x) = \log_a x$, $a > 0$	$f'(x) = 1/(x \log_e a)$

Exercises

ROUTINE

1. Find the first derivative of each of the following functions. Indicate any points at which the function is defined but at which the derivative is not. If a function is a composition of several functions, identify the functions of which it is the composition.

a) $f(x) = e^{x^2}$
b) $f(x) = (x^2 + 1)^{1/2}$
c) $g(t) = e^t/(t^2 + 1)$
d) $d(t) = (t^2 + 2t - 1)^3$
e) $s(x) = 3^x$
f) $k(t) = (1/2)^t - t$
g) $f(x) = (1/\sqrt{2\pi})e^{-x^2/2}$
h) $f(s) = (s^2 + 2s + 1)^4/(s^3 - 1)$
i) $q(w) = (((w + 1)^{1/2} + 1)^{1/2} + 1)^{1/2}$
j) $f(x) = (e^x - e^{-x})/(e^x + e^{-x})$
k) $g(x) = \log_e f(x)$

l) $g(x) = \log_a f(x)$
m) $h(t) = \log_e (t^2 + e^t)$
n) $f(x) = \log_3 (x^2 + e^{x^2})$
o) $g(x) = \log_e ((x - 1)/(x + 1))$
p) $f(x) = x^3/3$
q) $f(x) = x^4/4$
r) $f(x) = x^n/n$

CHALLENGING

2. Try to determine where each of the first derivatives found in Exercise 1 is positive in value; where it has a zero value; and where it has negative values.

3. For each function listed in Exercise 1, sketch the graph of the function together with the graph of its first derivative on the same set of axes. As in Exercise 3 of the preceding section, this may not be possible with our present techniques.

4. Answer each of the following questions, and indicate which rules analogous to those appearing in Table 4.1 are used in computing the answer to each question.
   a) A bug crawls along at 3 inches per minute. What is this rate in feet per minute?
   b) If the monthly sales of a company are 300 units in October, 300 in November, and 300 in December, what is the rate of change of sales?
   c) A man finds that his income from one source is increasing at the rate of 6 per cent per year, while income from another source is increasing at the rate of 8 per cent per year. If the two sources of income account for the man's total yearly income, and at the present time they are equal, at what rate is the man's yearly income increasing?

THEORETICAL

5. Suppose that $f(x) = m(n(p(x)))$, where m, n, and p are differentiable functions. Find an expression for f' in terms of m, n, p, and their derivatives. [Hint: Start by considering f as the composition of the two functions m and $n \circ p$; use the Chain Rule to calculate f' in terms of m, $n \circ p$, and their derivatives. Then apply the Chain Rule again to evaluate the derivative of $n \circ p$.]

6. The equation of a circle with center (0, 0) and radius r is

$$x^2 + y^2 = r^2,$$

or

$$y^2 = r^2 - x^2. \tag{i}$$

The value of y is dependent on the value of x; therefore, we may consider y to be a function of x. We can compute the derivatives of both sides of equation (i) with respect to x:

$$2yy' = -2x. \tag{ii}$$

a) Confirm the accuracy of equation (ii), and solve the equation for y'.
b) What is the slope of the line tangent to the circle at the point (x_0, y_0) of the circle? [Hint: Substitute x_0 and y_0 for x and y in the expression obtained for y', since y' is the slope of the circle thought of as the graph of y.]
c) Show that the slope obtained using y' for the slope of the line tangent to the circle at (x_0, y_0) is the same as the slope of the line tangent to the circle in the usual geometric sense. You will therefore be showing that, in this important case, both geometric and calculus tangents are the same line. You may use the facts that a tangent to a circle is perpendicular to the radius that passes through the point of tangency (Figure 4–10), and that if a line has slope m, then any line perpendicular to it has slope $-1/m$.

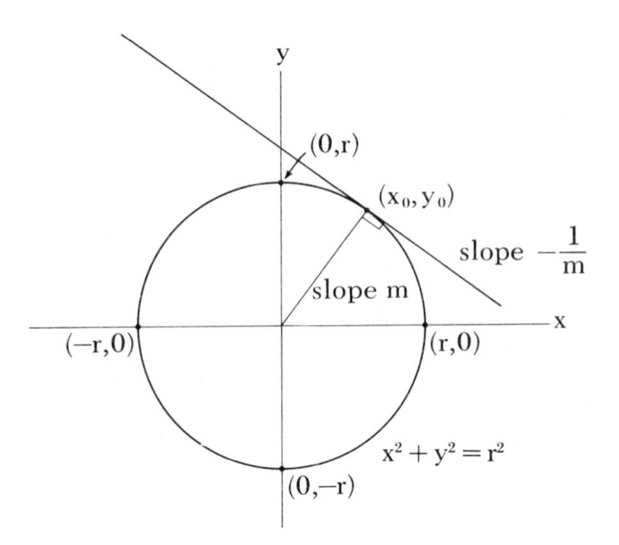

Figure 4–10

Review of Chapter 4

We define the *first derivative* $f'(a)$ of a function at a point a by

$$f'(a) = \lim_{h \to 0} \frac{f(a + h) - f(a)}{h}$$

if the limit on the right exists. The first derivative $f'(a)$ is the slope of the line tangent to the graph of f at the point $(a, f(a))$. If $f'(a)$ exists, we say that f is *differentiable* at a.

Given functions f and g, we define the *composition* $f \circ g$ of f with g by setting

$$(f \circ g) = f(g(x))$$

for each real number x for which both f and g are defined. If a function f is the composition of other functions or is formed using the operations introduced in Chapter 3, then the first derivative of f can often be expressed in terms of the component functions and their first derivatives. Rules for finding the derivative of a composite function in terms of the component functions are listed in Table 4.1. This table also includes the derivatives of the exponential and logarithmic functions.

REVIEW EXERCISES

1. Find the first derivative of each of the following functions, using any of the techniques introduced in this chapter.

a) $f(x) = -9x - 17$

b) $f(x) = 10^3$

c) $g(x) = 17x^2 + 21x + 0.45$

d) $h(x) = (x - 4)(6 - x)$

e) $f(x) = (x - 4)^4(6 - x)$

f) $g(x) = (x - 4)^4(6 - x)^{-2}$

g) $g(t) = t/(t^2 + 1)$

h) $h(t) = t^2 + t^{-2} + t^{0 \cdot 8} - 5 \log_e t$

i) $f(x) = (\log_{10} x)e^x$

j) $f(x) = 10^x + \log_e 10$

k) $f(x) = e^{7x^2 - 5x + 1}$

l) $f(x) = \log_e (3x^{-1} + 7 \log_e x + e^x)$

Vocabulary

increasing or decreasing function
relative maximum or minimum
n^{th} derivative
critical point
rate
differential equation
indefinite integral

Chapter 5

Some Uses of the Derivative

5.1 WHAT THE FIRST DERIVATIVE INDICATES ABOUT A FUNCTION

The first derivative f' of a function f enables us to gather information about f that generally would be difficult to obtain in any other way. Before indicating how the first derivative is used to solve practical problems, we raise the general question: What can we learn about f by studying f'?

Suppose that f is a function that is differentiable at a. Then $f'(a)$, as we saw in Section 4.1, is the slope of the line tangent to the graph of f at the point $(a, f(a))$. We may also think of this line as "approximating" the graph of f for values of x close to a (Figure 5–1). If $f'(a)$ is positive, then

TANGENT LINE TO THE GRAPH OF A FUNCTION DIFFERENTIABLE AT A POINT

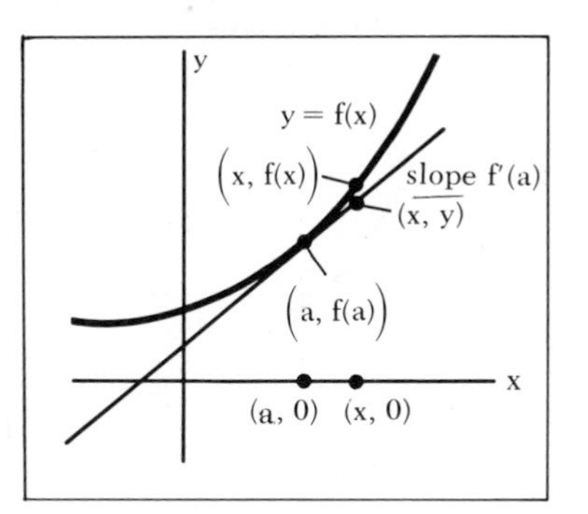

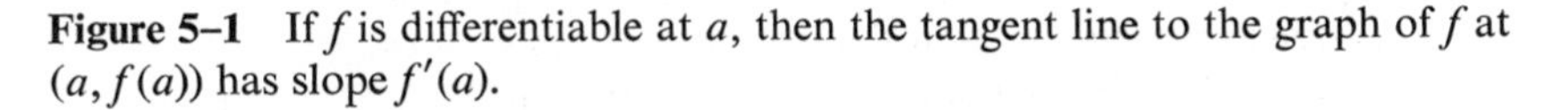

Figure 5–1 If f is differentiable at a, then the tangent line to the graph of f at $(a, f(a))$ has slope $f'(a)$.

the tangent line is "rising": the y-values of points on the tangent line increase as the x-values increase. Since the tangent line approximates the graph of f where x is close to a, we might then expect $f(x)$ to be increasing as x increases, provided that x is sufficiently close to a.

Example 1: Consider the function $f(x) = 3x^2 - 1$. For this function, $f'(x) = 6x$. Therefore, $f'(x)$ will be greater than 0 and the tangent line to the graph of f will have a positive slope if and only if x is greater than 0. It is then reasonable to conclude that the graph of f is rising ($f(x)$ increases as x increases) provided $x > 0$. In this case, our conclusion is easily confirmed by observation, for if x is positive and increasing, then $f(x) = 3x^2 - 1$ is also increasing. The graph of f appears in Figure 5–2.

GRAPH OF A FUNCTION AND ITS DERIVATIVE

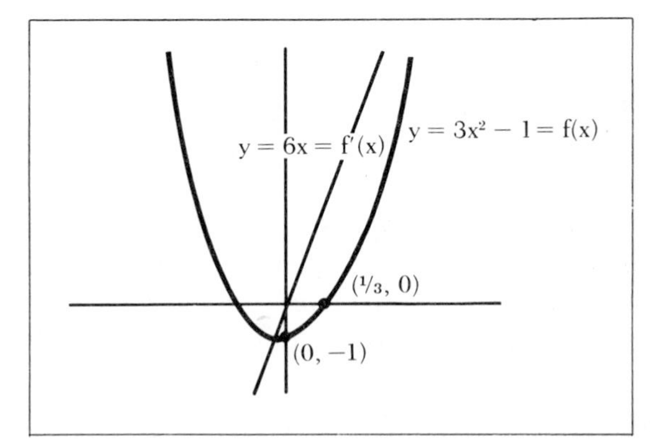

Figure 5–2 The graphs of $f(x) = 3x^2 - 1$ and $f'(x) = 6x$ are plotted together. When the graph of f' lies above the x-axis, then the graph of f is rising; when the graph of f' lies below the x-axis, the graph of f is falling.

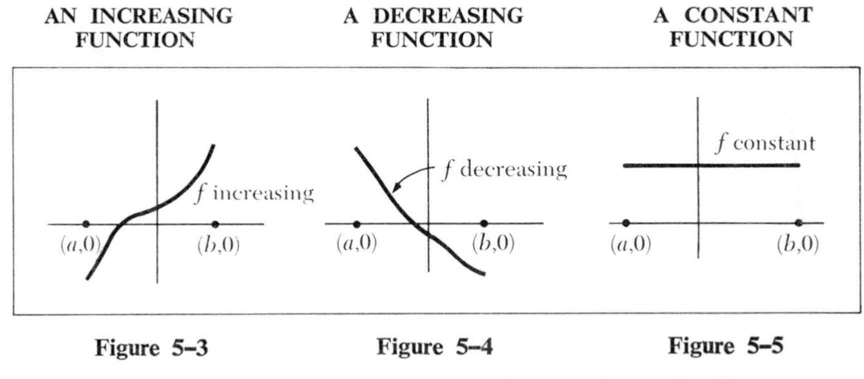

Figure 5–3 The function f is strictly increasing on (a, b).
Figure 5–4 The function f is strictly decreasing on (a, b).
Figure 5–5 The function f is constant on (a, b).

We might also note that if x is positive and increasing, then $f'(x) = 6x$, the slope of the line tangent to the graph of f, is also increasing. Since the larger the slope of a line, the more sharply the line is rising, we conclude that the graph of f itself rises more and more sharply as x becomes larger. This fact is also confirmed by inspection of f.

We observe further that if $x < 0$, then $f'(x) < 0$: the slope of the tangent line is negative and the graph of f, as we would expect, is falling ($f(x)$ is decreasing as x increases). When $x = 0, f'(x) = 0$, and the line tangent to the graph of f is horizontal. When $x = 0, f(x)$ has a minimum value. This, too, is not unexpected; since $f(x)$ is decreasing for $x < 0$ and increasing for $x > 0$, $f(0)$ must be the smallest value attained by f.

We now define more exactly some of the notions informally introduced above, and then we will formulate more precisely some of the rules for gaining information about f from the first derivative.

Definition 1: *A function f is said to be* ***strictly increasing*** *on an open interval (a, b) if f is defined at each point of (a, b) and if whenever x and x' are points of (a, b) with $x < x'$, then $f(x) < f(x')$ (Figure 5–3).*

A function f is said to be ***strictly decreasing*** *on an open interval (a, b) if f is defined at each point of (a, b) and if whenever x and x' are points of (a, b) with $x < x'$, then $f(x) > f(x')$ (Figure 5–4).*

We say that a function f is ***constant*** *on an open interval (a, b) if $f(x) = k$, where k is some constant, for every x in (a, b) (Figure 5–5).*

A function is said to have a ***relative maximum*** *at c if $f(c)$ is the largest function value that f assumes in some open interval containing c; in other words, for all x for which f is defined in some open interval containing c, $f(x) \leq f(c)$ (Figure 5–6). Similarly, a function f is said to have a* **relative maximum** *at c if, for all x for which f is defined in some open interval containing c, $f(x) \geq f(c)$ (Figure 5–7).*

MAXIMUM AND MINIMUM FUNCTION VALUES IN AN OPEN INTERVAL

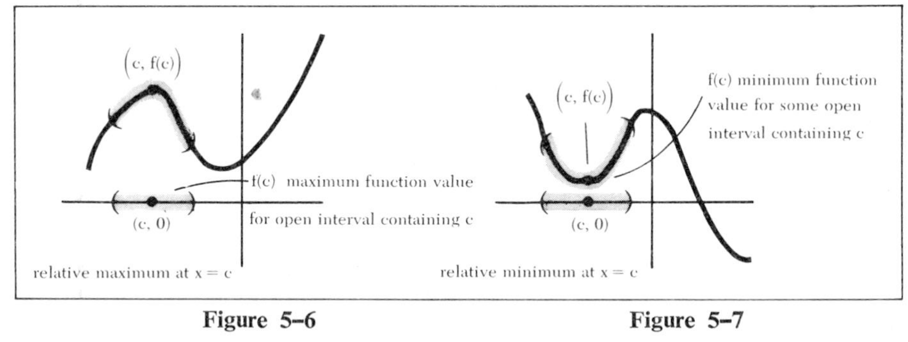

Figure 5–6 **Figure 5–7**

Figure 5–6 A relative maximum on (a, b).
Figure 5–7 A relative minimum on (a, b).

Proposition 1: If f is differentiable at each point of the open interval (a, b) and $f'(x) > 0$ for each x in (a, b), then f is strictly increasing on (a, b). If f is differentiable at each point of an open interval (a, b) and $f'(x) < 0$ for each x in (a, b), then f is strictly decreasing on (a, b).

We omit a formal proof of this result, but we should be convinced that it is entirely in accord with what we expect of the first derivative. If each line tangent to the graph of $y = f(x)$ with x in (a, b) has a positive slope, then the values of $f(x)$ ought to be increasing as x increases in (a, b). If each tangent line has a negative slope, then the function values should be decreasing.

Suppose now that f is a function that is differentiable at each point of the open interval (a, b), and suppose that c is a point of (a, b) at which f has a relative maximum. By choosing (a, b) appropriately—it would be the open interval about c mentioned in the definition of a relative maximum—we may assume that $f(c)$ is the largest function value of f on (a, b) (Figure 5–8). Since $f(c)$ is the largest function value for x in (a, b), we see that $f(x)$ must increase to $f(c)$ as x approaches c, provided x is close to c. The function values $f(x)$ cannot decrease to $f(c)$, for then there would be values of $f(x)$ on (a, b) greater than $f(c)$. Likewise, $f(x)$ must

A RELATIVE MAXIMUM AT $x = c$

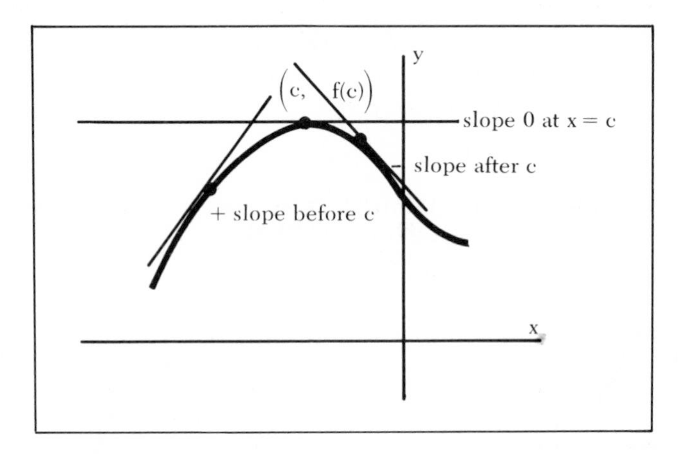

Figure 5–8 The tangent line to the graph of f at $(c, f(c))$ will be horizontal if f is defined in an open interval containing c and f has a relative maximum at $x = c$. The graph of f will be rising to the left of $(c, f(c))$ and falling to the right of $(c, f(c))$; thus, $f'(x)$ will be non-negative if x is close to but less than c, and $f'(x)$ will be non-positive if x is close to but greater than c.

start to decrease once x is past c, for if $f(x)$ kept increasing after x passed c, then again $f(c)$ could not be the largest function value for x in (a, b). Thus we would expect the slope of f to be non-negative just to the left of c and non-positive just to the right of c. In other words, we expect $f'(x) \geq 0$ for $x < c$ and $f'(x) \leq 0$ for $x > c$, provided x is suitably close to c; hence at c we can reasonably expect $f'(c) = 0$. The preceding discussion is intended as a basis for (but not a proof of) the following proposition.

Proposition 2: If f is differentiable at c and f has either a relative maximum or a relative minimum at c, then $f'(c) = 0$.

We then see that the first derivative not only yields information about where a function is increasing or decreasing, but it also may help us find relative maxima and minima of the function. We say "may help us find" rather than "always enables us to find," because even if $f'(c) = 0$ it is not certain that f has either a relative maximum or a relative minimum. Moreover, maxima and minima sometimes occur where we cannot find them by using calculus. We will return to this problem later. The following examples are meant to illustrate Propositions 1 and 2.

Example 2: Let $f(x) = 1/(x^2 + 1) = (x^2 + 1)^{-1}$. Then $f'(x) = -(x^2 + 1)^{-2}(2x)$, or

$$f'(x) = \frac{-2x}{(x^2 + 1)^2}\,. \tag{1}$$

Since $(x^2 + 1)^2$ is always positive, the sign of equation (1) will be the same as the sign of $-2x$. Clearly, $-2x$ will be positive when x is negative, and it will be negative when x is positive. We therefore conclude that

$$\begin{aligned} f'(x) > 0 \quad &\text{if} \quad x < 0, \\ f'(x) < 0 \quad &\text{if} \quad x > 0. \end{aligned} \tag{2}$$

Note also that $f'(0) = 0$; therefore it is possible that a relative maximum or minimum exists at $x = 0$. From inequality (2) we conclude that f is increasing for $x < 0$ and decreasing for $x > 0$; from this it follows that $f(0)$ is the maximum value for f. For this function, we may note that the open interval on which f is defined, and for which $f(0)$ is a relative maximum, is the entire set of real numbers. Since we are thus assured that $f(x) < f(0)$ for all x, we say that $f(0)$ is the *absolute maximum* for f.

In this instance, we have obtained no new information about f by examining f'. Nevertheless, we confirmed what we had already discovered without the use of the calculus, thereby strengthening our confidence that the definition of the derivative given in Chapter 4 adequately embodies the intuitive notions underlying it. We will see later that the calculus can provide information beyond that which can be obtained simply by inspection of the definition of f.

Example 3: Let $f(x) = (x - 1)(x - 2) = x^2 - 3x + 2$. Then

$$f'(x) = 2x - 3. \tag{3}$$

The graph of f' appears in Figure 5–9. We see that $f'(x) > 0$ for $x > 3/2$, that $f'(x) < 0$ for $x < 3/2$, and that $f'(3/2) = 0$. It then follows that f is decreasing for $x < 3/2$, increasing for $x > 3/2$, and has a minimum value when $x = 3/2$. As x increases, the slope of f is increasing; this is seen geometrically in the rising graph of f'. This means that for $x < 3/2$, f decreases less and less rapidly as x increases. However, once x becomes greater than $3/2$, f increases more and more rapidly as x increases. The function f has the value 0 when $x = 1$ and when $x = 2$; $f(3/2)$, its minimum value, is equal to $-1/4$. Combining the information we have obtained about f, we can sketch an accurate graph of it (Figure 5–10).

The graph of the function in Example 3 is a *parabola*. Any equation of the form $y = ax^2 + bx + c$, where a, b, and c are constants and $a \neq 0$, has a graph that is a parabola. If

$$f(x) = ax^2 + bx + c, \tag{4}$$

then

$$f'(x) = 2ax + b. \tag{5}$$

The graph of f' will have either the form shown in Figure 5–11 or that in Figure 5–12, depending on whether a is positive or negative. If a is positive, then f will have a minimum at $x = -b/2a$; if a is negative, then f has a maximum at $x = -b/2a$. The zeros of f, those values of x for which $f(x) = 0$, are found by solving the quadratic equation

$$ax^2 + bx + c = 0. \tag{6}$$

We can solve equation (6) by factoring or by use of the so-called *quadratic formula*. According to the quadratic formula, which we state without proof, the roots of equation (6) are

$$x_1 = \frac{-b - \sqrt{b^2 - 4ac}}{2a}$$

THE GRAPHS OF A FUNCTION AND ITS DERIVATIVE

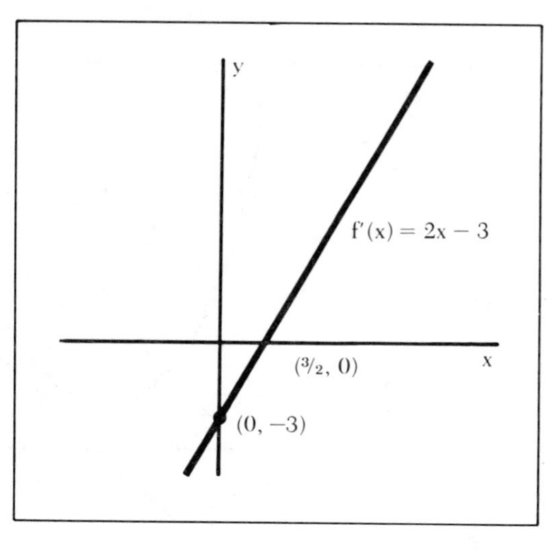

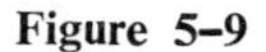
Figure 5–9

Figures 5–9 and **5–10** The graph of f will be falling where the graph of f' lies below the x-axis and rising where the graph of f' lies above the x-axis. We can read from this graph that $f(x) = x^2 - 3x + 2$ is a function which decreases less and less sharply for $x < 3/2$, has a minimum at $x = 3/2$ (where $f'(3/2) = 2x - 3 = 0$), and increases more and more sharply for $x > 3/2$. This information, together with the points at which the graph of f crosses the x- and y-axes, enables us to obtain an accurate sketch of the graph of f.

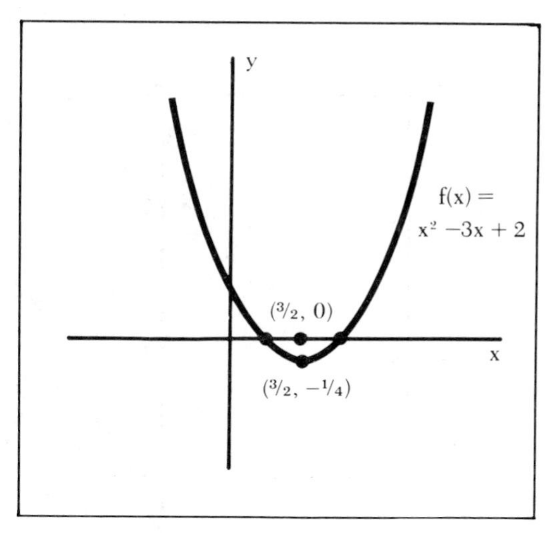

Figure 5–10

and

$$x_2 = \frac{-b + \sqrt{b^2 - 4ac}}{2a}$$

If $b^2 - 4ac < 0$, which means that the roots of equation (6) are imaginary numbers, then the graph of f will not cross the x-axis, and $f(x)$ cannot be 0

AN AID TO GRAPHING A QUADRATIC FUNCTION

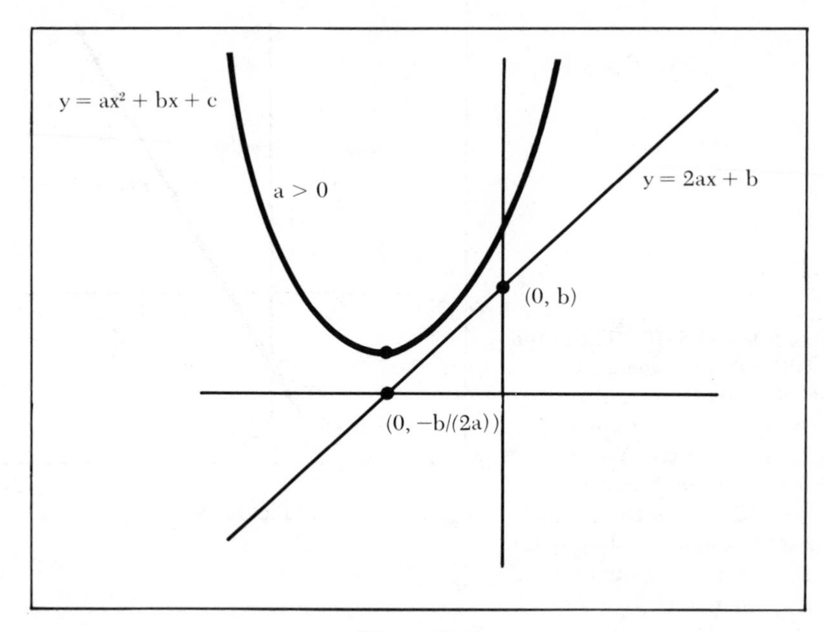

Figure 5–11

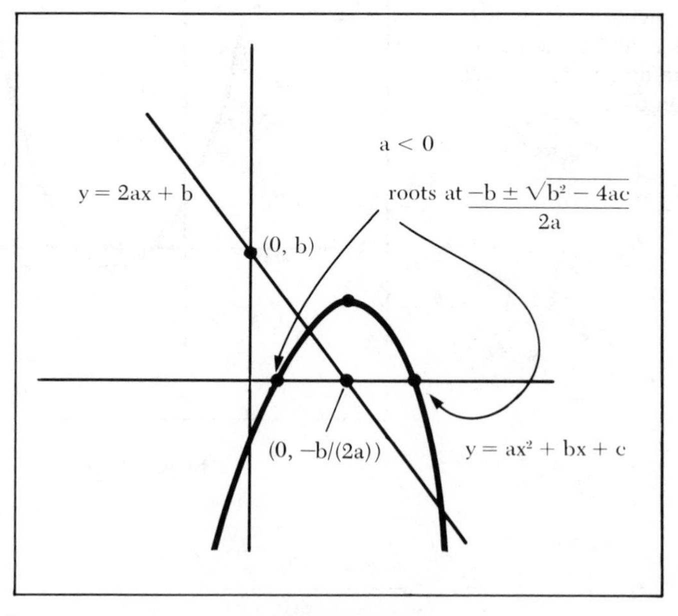

Figure 5–12

Figures 5–11 and **5–12** If $a > 0$, then the graph opens upward and has a minimum at $x = -b/2a$; if $a < 0$, then the graph opens downward and has a maximum at $x = -b/2a$.

for any real number. Possible graphs of f are sketched along with f' in Figures 5–11 and 5–12.

In the next section we continue our discussion of the information that the first derivative can give us about a function.

Exercises

ROUTINE

1. Find the first derivative of each of the following functions. Then find the zeros of each first derivative.

a) $f(x) = -x^2 - x - 1$
b) $f(x) = 3x^2 + 6x + 9$
c) $g(t) = 1/t^2$
d) $f(t) = (1 + t)/(1 - t)$
e) $f(x) = e^{x^2}$
f) $f(x) = e^{-x^2}$
g) $h(t) = te^{-t^2}$
h) $f(x) = \log_e x^2$
i) $f(x) = x \log_e x$
j) $g(x) = x^2 \log_3 x$
k) $h(x) = 1/(e^x + 1)$
l) $f(x) = \log_4 (xe^x)$

2. For each function in Exercise 1, determine for which real numbers the first derivative has positive values, and for which real numbers the first derivative has negative values. This will tell us where the function is strictly increasing and where it is strictly decreasing (see Proposition 1).

3. Use the information gained in Exercise 2 to determine the nature of each of the zeros of the first derivatives found in Exercise 1: determine whether the function has a relative maximum, relative minimum, or neither a relative maximum nor minimum at each zero of its first derivative.

4. Use the information gained in Exercises 1 through 3 to sketch the graphs of the functions given in Exercise 1.

CHALLENGING

5. The graphs in Figure 5–13 are the sales curves of three new companies, showing the monthly sales of each company over the same period of time.

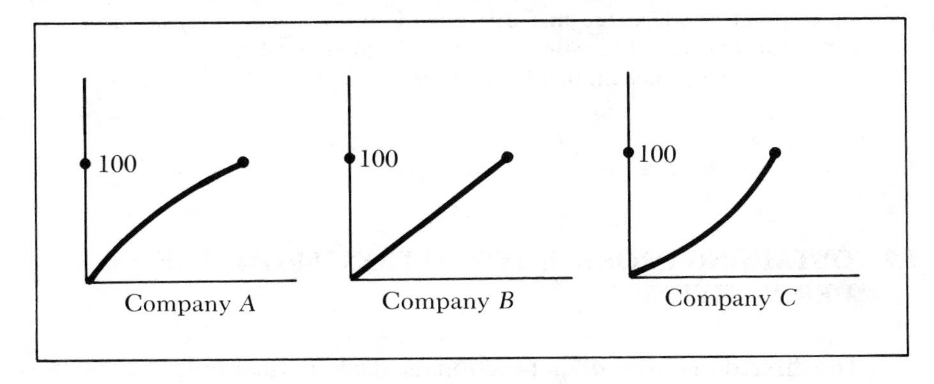

Figure 5–13 Exercise 5.

Each company is successful and sells more each month than in the previous month, but which of the companies has the most promising rate of increase of sales? What might be a quantitative measure of this rate of increase?

THEORETICAL

6. A formal proof of Proposition 2 proceeds as follows (supply the details): Suppose f is differentiable and has a relative maximum at $x = c$. Consider

$$\frac{f(c + h) - f(c)}{h}. \tag{7}$$

The limit of expression (7) as h approaches 0 is $f'(c)$.

a) If h is negative and close to 0, what sign does the numerator of expression (7) have? What sign does expression (7) have?
b) From the information gained in part (a), why must we conclude that $f'(c)$ is no greater than 0?
c) What sign does expression (7) have if h is positive and close to 0? Keep in mind that $f(c)$ is the largest function value in some open interval that contains c.
d) Why can we conclude from part (c) that $f'(c)$ is no less than 0? Since $f'(c)$ is no greater than 0 but is also no less than 0, we conclude that $f'(c) = 0$.

7. The function $f(x) = |x|$ has a minimum value at $x = 0$. Explain why this minimum value cannot be detected using Proposition 2.

8. Consider the function $g(x) = x$ defined for x in the closed interval $[0, 1]$. With x restricted to the domain of g, what is the minimum value of $g(x)$? What is the maximum value of $g(x)$? Can either the maximum or minimum value be detected using Proposition 2? Explain.

9. Let f be a function that is differentiable at every real number. Justify the following statement: Between any two zeros of f there must be a point at which the first derivative is 0. You may use the fact that a function that is continuous on a closed interval has a maximum and minimum value on that interval.

5.2 OBTAINING MORE INFORMATION FROM THE DERIVATIVE

The first derivative of a function is itself a function; hence it is often feasible to take the first derivative of the first derivative.

Definition 2: *Suppose that f is a differentiable function. Then $(f')'$, the first derivative of the first derivative of f, is called the* ***second derivative*** *of f. We generally denote the second derivative by f''.*

One can also take the first derivative of f''; we call $(f'')'$ the ***third derivative*** *of f. This third derivative may be denoted by f''' or by $f^{(3)}$. In general, the* ***n^{th} derivative*** *of f is the first derivative of the $(n-1)^{th}$ derivative, and is denoted by $f^{(n)}$.*

Example 4: Let $f(x) = 4x^3 - 7x^2 + 3x - 1$. Then

$$f'(x) = 12x^2 - 14x + 3,$$

$$f''(x) = 24x - 14,$$

$$f^{(3)}(x) = 24,$$

and

$$f^{(4)}(x) = 0.$$

All derivatives of f higher than the fourth are also equal to zero.

We can use f'' to gain information about f' in the same way that we use f' to gain information about f. From f'' we obtain information about f' and also, indirectly, about f. For example, suppose that $f''(x)$ is positive for every x in the open interval (a, b). This means that f' is an increasing function on (a, b). In this case, then, if $f'(x)$ itself is positive for x in (a, b), then f' is both positive and increasing on (a, b). Therefore, the slope of f (the rate at which f is increasing) is positive and increasing, and the graph of f rises more and more sharply as x increases from a to b. If, however, $f'(x)$ is negative for each x in (a, b), then the graph of f falls less and less sharply as x increases from a to b. In Figures 5–14 to 5–18 we illustrate graphically the relationship between f', f'', and the graph of f. The conditions stated with each figure are for x in (a, b).

We saw earlier that if a function f has a relative maximum or minimum at $x = c$ and f is differentiable at c, then $f'(c) = 0$. It is also possible to have $f'(c) = 0$ even when f has neither a maximum nor a minimum at c. Such is the case with $f(x) = x^3$ at $x = 0$. Even if we are certain that f has either a maximum or a minimum at c, we still must decide which of these alternatives is in fact the case. We will soon see that the second derivative often furnishes a ready test for the nature of c. To facilitate this discussion we introduce a term for a point at which the first derivative is 0.

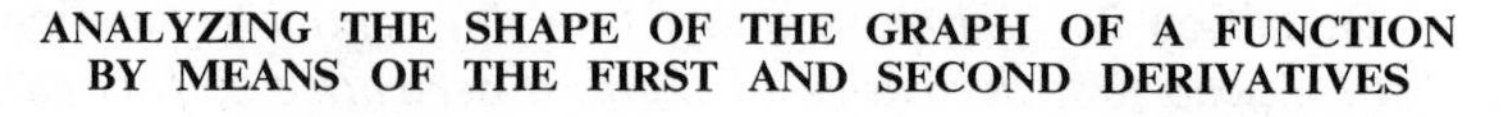

ANALYZING THE SHAPE OF THE GRAPH OF A FUNCTION BY MEANS OF THE FIRST AND SECOND DERIVATIVES

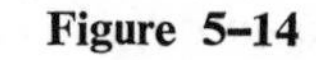

Figure 5–14 **Figure 5–15**

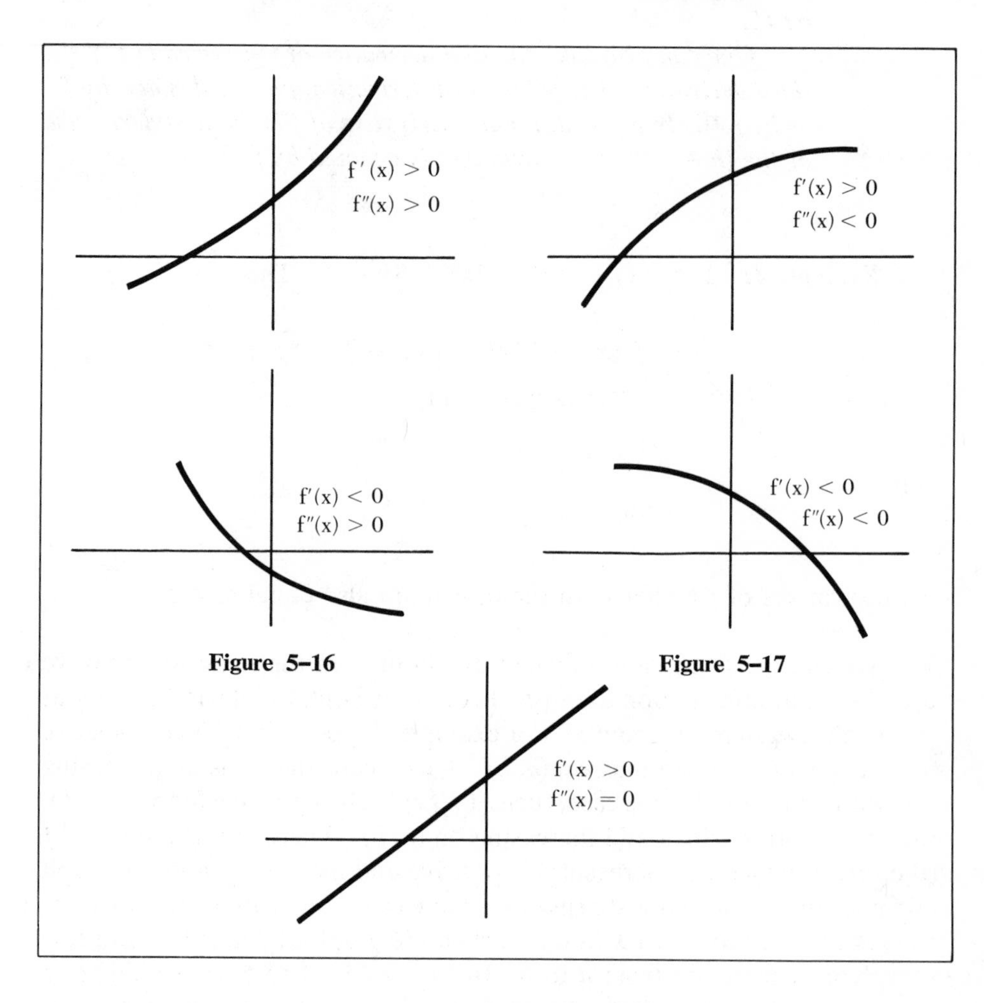

Figure 5–18

Figures 5–14, 5–15, 5–16, 5–17, and **5–18** The second derivative of f tells us whether the first derivative of f is increasing ($f''(x) > 0$), decreasing ($f''(x) < 0$), or constant ($f''(x) = 0$). Thus the first and second derivatives together tell us more about the behavior of f than does the first derivative alone.

Definition 3: *If f is a function and c is a number such that $f'(c) = 0$, then we call c a* ***critical point*** *of f.*

The problem then is to identify the nature of the critical points of f. Suppose f has a relative maximum at a critical point c. Then as x approaches c for $x < c$, $f(x)$ is increasing; once $x > c$, then $f(x)$ decreases as x increases, at least for x close to c (Figure 5–19). Thus $f'(x)$ will be non-negative for x close to c and $x < c$, and $f'(x)$ will be non-positive for

ANALYZING CRITICAL POINTS BY MEANS OF THE SECOND DERIVATIVE

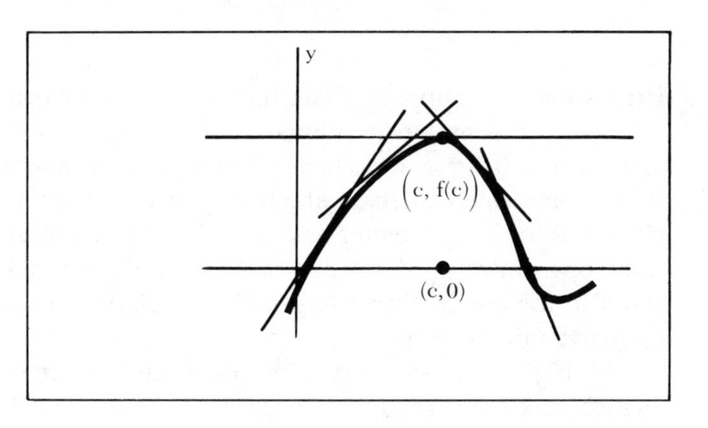

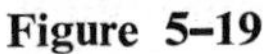
Figure 5–19

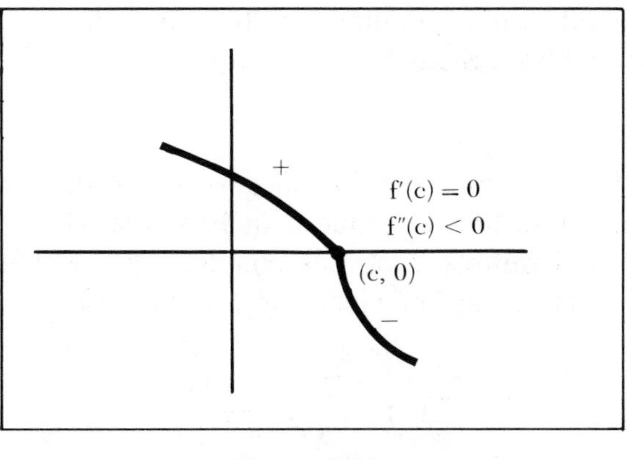

Figure 5–20

Figures 5–19 and **5–20** If f has a critical point at $x = c$, f'' is continuous at $x = c$, and $f''(c) < 0$, then $f'(c) = 0$ and f' is decreasing near c; hence $f'(x)$ must be positive for x close to and less than c, and $f'(x)$ must be negative for x close to and greater than c. This implies that f has a relative maximum at c.

x close to c and $c < x$. If we slide a tangent line along the graph of f near $(c, f(c))$ (Figure 5–19), it becomes apparent that $f'(x)$ must be decreasing for x close to c. Note that $f'(x)$ in fact passes from positive to negative. If f' is decreasing in the vicinity of $x = c$, then f'', which measures the rate of change of f', should be negative in that same region.

Now suppose it is given that $f'(c) = 0$ and $f''(c) < 0$. This implies that f' is decreasing at c and $f'(x)$ must be positive for $x < c$ and negative for $x > c$ (Figure 5–20). Thus f is increasing for $x < c$ and decreasing for $x > c$, so we are led to conclude that f has a relative maximum at c. An analogous discussion would help show that if $f'(c) = 0$ and $f''(c) > 0$, then f has a relative minimum at c. The following proposition states this result more precisely.

Proposition 3: Suppose f is a function with a critical point c.

a) If, in some open interval about c, $f'(x) < 0$ for $x < c$ and $f'(x) > 0$ for $c < x$, then f has a relative minimum at c. If, in some open interval about c, $f'(x) > 0$ for $x < c$ and $f'(x) < 0$ for $c < x$, then f has a relative maximum at c. If, in some open interval about c, $f'(x) < 0$ or $f'(x) > 0$ for $x < c$ and also for $c < x$, then f has neither a relative maximum nor a relative minimum at c.

b) If f'' is continuous at c and $f''(c) < 0$, then f has a relative maximum at c. If f'' is continuous at c and $f''(c) > 0$, then f has a relative minimum at c. (The continuity of f'' is necessary to ensure that $f''(x)$ has the same sign as $f''(c)$ in some open interval containing c.) If $f''(c) = 0$, we gain no information about the behavior of f at c.

Example 5: We continue the discussion of the function $f(x) = 1/(x^2 + 1)$, which was first introduced in Example 34 of Chapter 1 and was continued in Example 2 of this chapter. We will use the second derivative of f to refine our knowledge of f. Since $f'(x) = -2x/(x^2 + 1)^2$, we find that

$$f''(x) = \frac{2(3x^2 - 1)}{(x^2 + 1)^3}. \tag{8}$$

We note from equation (8) that f' itself has critical points (points at which f'' is 0) at $1/\sqrt{3}$ and $-1/\sqrt{3}$. We have already seen that $f'(x) > 0$ and f is increasing for $x < 0$. However, we can readily show that f' has a relative maximum at $-1/\sqrt{3}$; hence, when $x = -1/\sqrt{3}$, f is increasing most rapidly. Furthermore, although f is decreasing for $0 < x$, f' has a relative minimum at $1/\sqrt{3}$; hence, when $x = 1/\sqrt{3}$, f is decreasing most rapidly. We can gather the following information from equation (8):

If $x < -1/\sqrt{3}$, then $f''(x) > 0$, and thus f' is increasing.

If $-1/\sqrt{3} < x < 1/\sqrt{3}$, then $f''(x) < 0$, and thus f' is decreasing.

If $x > 1/\sqrt{3}$, then $f''(x) > 0$, and thus f' is again increasing.

We previously found that f has a critical point at $x = 0$. Since f'' is continuous at 0 and $f''(0) = -2$, we confirm our earlier conclusion that f has a relative maximum at $x = 0$. All the information about f gained from our inspection of the definition of f, as well as from f' and f'', is now compiled in a more accurate graph than would be possible by inspection or use of the first derivative alone. The graph of f is shown in Figure 5–21.

ANOTHER LOOK AT THE FUNCTION $f(x) = 1/(x^2 + 1)$

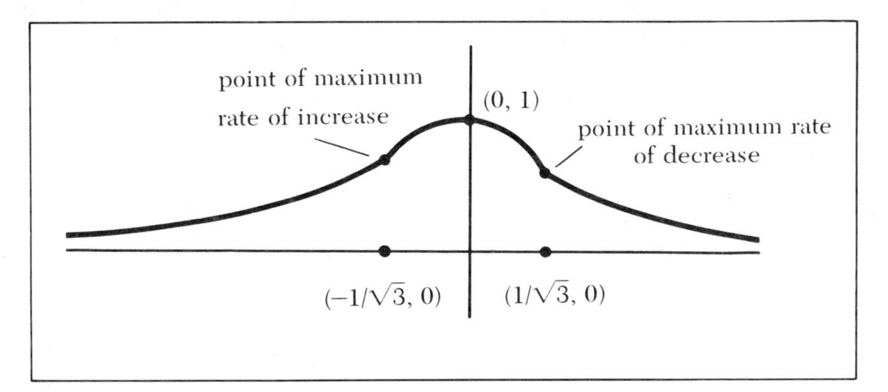

Figure 5–21 We not only can determine where the function is increasing or decreasing, but also can determine how fast it is increasing or decreasing and where the points of maximum rate of increase or decrease occur.

Example 6: We now consider the function $f(x) = xe^x$. First we compute

$$f'(x) = xe^x + e^x = e^x(x + 1), \tag{9}$$

and

$$f''(x) = e^x(1) + e^x(x + 1) = e^x(x + 2). \tag{10}$$

We note that f has a critical point at $x = -1$. Since $f''(-1) = e^{-1} > 0$, f has a relative minimum at -1. By inspection we see that $f'(x) > 0$ for $-1 < x$ and $f'(x) < 0$ for $x < -1$; therefore, f is decreasing for $x < -1$ and increasing for $-1 < x$.

From equation (10) we see that f' has a critical point at -2. Since $f''(x) < 0$ for $x < -2$ and $f''(x) > 0$ for $-2 < x$, we find that f' has a relative minimum at -2. The slope of f decreases until it reaches a minimum at $x = -2$, and then it increases for $x > -2$. We should keep in mind that the fact that the slope is increasing does not necessarily imply that it is positive.

We now make some direct observations about the definition of f. First, since $e^x > 0$, $f(x) < 0$ when $x < 0$, and $f(x) > 0$ when $x > 0$. When $x > 0, f(x)$ will be increasing even faster than e^x.

What happens to $f(x)$ when x is negative and decreasing? Note that as x goes further and further along the negative x-axis, e^x approaches 0 while x becomes larger and larger in absolute value, and so inspection alone cannot help us predict what happens to xe^x.

However, the minimum value of f is $f(-1) = -e^{-1} = -1/e$. For $x < -1, f$ has negative slope; hence for $x < -1$ the values of $f(x)$ are negative and decreasing as x approaches -1. This strongly implies—although we have no proof—that $f(x)$ tends to 0 as x moves further and further to the left on the negative x-axis. Using the information obtained above, we now sketch the graph of f (Figure 5–22).

THE GRAPH OF $f(x) = xe^x$

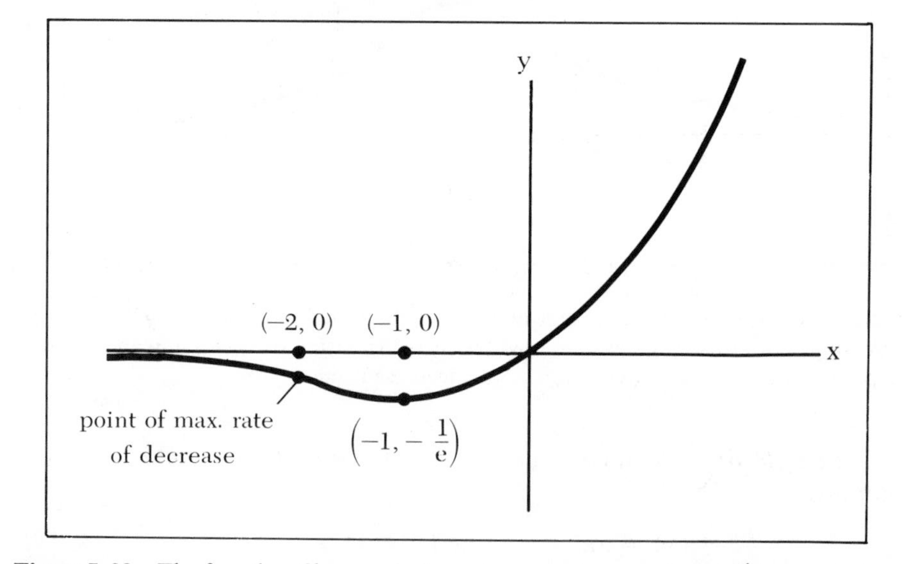

Figure 5–22 The function f has a relative minimum at $x = -1$ and f' has a minimum at $x = -2$. When x is negative and decreasing, the graph has the x-axis as a limit.

We close this section with a basic fact about differentiable functions that will be vital in our study of the definite integral.

Suppose that a function f is defined and continuous on a closed interval $[a, b]$ and is differentiable at every point of the open interval (a, b). Consider the graph of f for the closed interval $[a, b]$ (Figure 5–23).

THE MEAN VALUE THEOREM

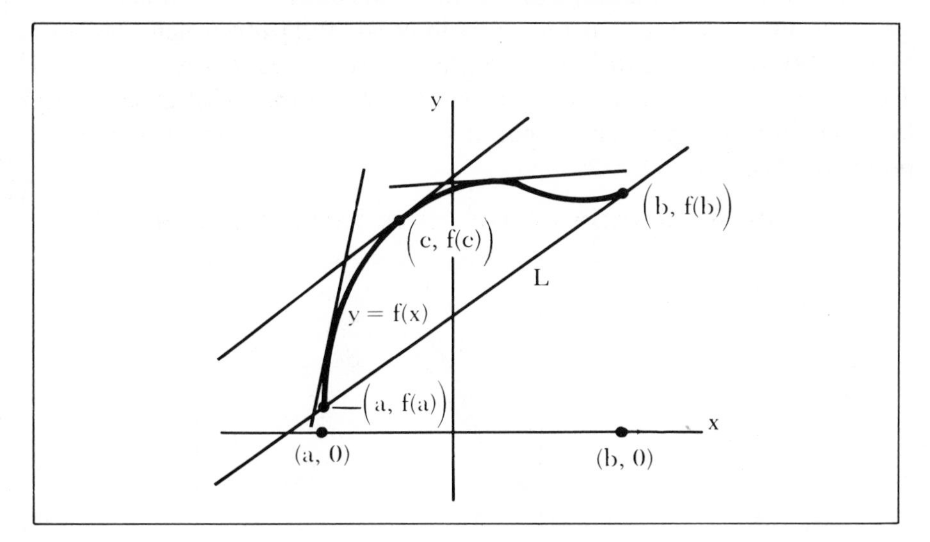

Figure 5–23 If f is defined and continuous on $[a, b]$ and differentiable on (a, b), then the tangent line to the graph at some point between $(a, f(a))$ and $(b, f(b))$ will be parallel to the line determined by $(a, f(a))$ and $(b, f(b))$.

The slope of the line determined by the endpoints of this graph is

$$\frac{f(b) - f(a)}{b - a} \tag{11}$$

It seems reasonable that if we guide a line along the graph of f, keeping the line tangent to the graph, then at some point $(c, f(c))$ the tangent line will be parallel to the line determined by $(a, f(a))$ and $(b, f(b))$, the endpoints of the graph, and will have the slope given by expression (11). The slope of the tangent line is, of course, $f'(c)$. We can therefore state the following proposition.

Proposition 4 (The Mean Value Theorem): If a function f is continuous on $[a, b]$ and differentiable at each point of (a, b), then there is at least one point c in (a, b) such that

$$f'(c) = \frac{f(b) - f(a)}{b - a}. \tag{12}$$

Example 7: The function $f(x) = 3x^2 - 2x + 2$ is continuous on $[0, 1]$ and differentiable at each point of $(0, 1)$. The graph of $y = f(x)$ for x in $[0, 1]$ is given in Figure 5–24. The endpoints of this graph are $(0, 2)$ and $(1, 3)$, and the slope of the line connecting these points is 1. We will now find a point c in the open interval $(0, 1)$ for which $f'(c) = 1$. Since $f'(x) = 6x - 2$, to find c we must solve the equation $6c - 2 = 1$. Therefore $c = 1/2$. The line tangent to the graph of f at the point $(1/2, f(1/2))$ has slope 1 and is parallel to the line determined by the endpoints $(0, 2)$ and $(1, 3)$.

AN APPLICATION OF THE MEAN VALUE THEOREM

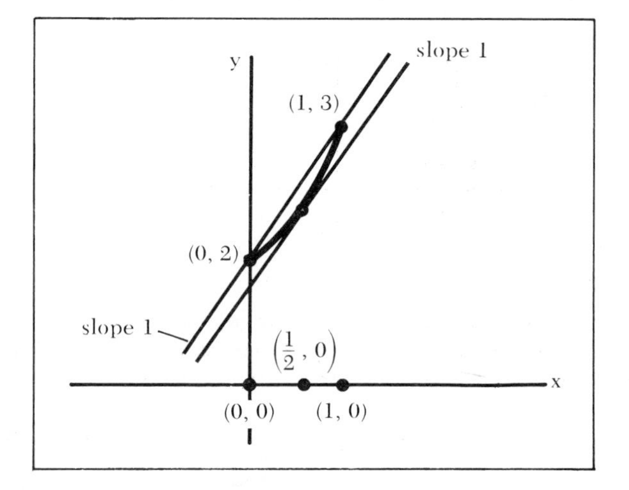

Figure 5–24 Taking the function $f(x) = 3x^2 - 3x + 2$ defined on $[0, 1]$, the line tangent to the graph of f at $(1/2, f(1/2))$ is parallel to the line determined by $(0, 2)$ and $(1, 3)$.

Exercises

ROUTINE

1. With the help of the first and second derivatives, sketch the graph of each of the following functions. Label any points on the graph where the first derivative of the function has a relative maximum or minimum.

a) $f(x) = x^2 + 4x - 9$
b) $g(x) = x^3 - 3x^2 + 4x + 7$
c) $f(x) = xe^{-x}$
d) $f(t) = -3t^3 - 4t^2 + t$
e) $s(t) = e^t/t$
f) $f(x) = e^x/(x^2 + 1)$
g) $f(x) = e^x x^{1/2}$
h) $f(x) = xe^{-4x}$
i) $f(x) = x^4 + 5x^3 + 6x^2 + x - 9$
j) $g(x) = (x^2 - 1)(x + 3)(x + 9)$
k) $f(x) = (e^x + 1)^4$
l) $f(t) = (1/(1 + t^2))^3$
m) $f(x) = (x^2 - 1)/(x^2 + 1)$
n) $g(x) = x + x^{-1} + x^{-2}$
o) $f(t) = ((t + 1)^2 + 1)^3$
p) $g(x) = x/(x^2 + 3x + 5)$
q) $f(x) = e^{(x^2+3x+2)}$
r) $f(x) = \log_e (x^2 + 1)$
s) $g(x) = \log_3 ((x - 1)/(x + 1))$
t) $f(x) = \log_e (x + e^x)$

2. Sketch the graphs of the first derivative of each function given in Exercise 1, if possible, on the same set of axes as the graph of the function, to see the relationship between the graphs. Also try to graph the second derivative in at least two cases.

3. Indicate those portions of the graphs of the functions in Exercise 1 where the function is increasing but its first derivative is decreasing; also indicate where the function is decreasing but the first derivative is increasing.

4. A function that is continuous on a closed interval always has a maximum value and a minimum value on that closed interval. Find the maximum and minimum values of the given functions on the closed intervals indicated.

a) $f(x) = 2x^3 + 3x - 1$; $[-3, 3]$
b) $g(x) = x^2$; $[0, 1]$
c) $x(t) = t^2 + (4/t^2)$; $[-2, 2]$
[Note: Is this function continuous on the given closed interval? Does it have a maximum value on this interval?]
d) $h(x) = (x + 2)^2(x - 3)^3$; $[-3, 4]$

THEORETICAL

5. The Mean Value Theorem can be used to prove that a function is increasing wherever its first derivative is positive. We now sketch such a proof; you should supply any necessary details. Suppose that a function f is differentiable at each point of the open interval (a, b), and $f'(x) > 0$ for each x in (a, b). Suppose that x and c are points of (a, b) with $x < c$. The hypotheses of Proposition 4 apply to f with respect to $[x, c]$. Why? Thus there is a point y of (x, c) such that $f'(y) = (f(c) - f(x))/(c - x)$, or

$$f'(y)(c - x) = f(c) - f(x). \tag{13}$$

What is the sign of $f'(y)$? What is the sign of $c - x$? What, then, is the sign of $f(c) - f(x)$? We can therefore conclude that $f(x) < f(c)$; hence f is increasing on (a, b).

6. Use the Mean Value Theorem to prove that if the first derivative of a function is 0 at each point of an open interval (a, b), then the function is constant on (a, b). [Hint: Use an argument similar to that of Exercise 5 to show that $f(c) - f(x) = 0$ for any points c and x of (a, b).]

7. Use the result of Exercise 6 to prove that if two functions have the same first derivative on an open interval (a, b), then they differ by a constant factor on (a, b). [Hint: Suppose that $f' = g'$; then $(f - g)'(x) = f'(x) - g'(x) = 0$ for all x in (a, b). Now apply Exercise 6 to $f - g$.]

5.3 SOME PRACTICAL APPLICATIONS OF THE DERIVATIVE

Maximum-Minimum Problems

In real life, many instances occur in which the most advantageous value of some variable or function is sought. For example, a manufacturer may wish to know how many of a certain item he should produce to maximize his profits; an airline may want to know the speed at which a plane should cruise to maximize the airline's profits; a container company may want to know the dimensions of the box of largest volume that can be produced if the box is to satisfy certain specifications. We are often able to translate such problems into ones of finding maximum or minimum values of some differentiable function, and so we may use the methods of solution developed in the previous two sections. We now present a classical problem in maximizing a given function.

Example 8: A man has a rectangular piece of cardboard measuring 12 inches by 24 inches. What is the volume of the largest box (without top) that the man can make by cutting the same sized squares from each corner of the cardboard (Figure 5–25) and folding up the remaining edges? By cutting away squares with side x, the man will obtain a box with dimensions x by $24 - 2x$ by $12 - 2x$ inches. Thus the volume $V(x)$ of the box is

$$\begin{aligned} V(x) &= x(24 - 2x)(12 - 2x) \\ &= 4x^3 - 72x^2 + 288x \text{ (cubic inches).} \end{aligned} \tag{14}$$

The smallest possible value of x that is consistent with the physical realities of the problem is 0, while the largest possible value is 6. Therefore, we restrict our attention to the closed interval $[0, 6]$, and we ask: For what value of x in $[0, 6]$ is $V(x)$ maximal? In Figure 5–26 we can see how $V(x)$ behaves in relation to x for values of x in $[0, 6.]$

We now find the critical points of V. Note first that V is differentiable at each point of $(0, 6)$. If V has a relative maximum at some point of $(0, 6)$, then its first derivative will be 0 at that point. We differentiate $V(x)$ and obtain

$$V'(x) = 12x^2 - 144x + 288. \tag{15}$$

Solving the equation $12x^2 - 144x + 288 = 0$ (or $x^2 - 12x + 24 = 0$) by means of the quadratic formula (cf. Section 5.1), we find that V has critical points at $x = 6 + 2\sqrt{3}$ and $x = 6 - 2\sqrt{3}$. We immediately discard the point $x = 6 + 2\sqrt{3}$ since it is clearly outside the interval $[0, 6]$. On the other hand, $6 - 2\sqrt{3}$ is not only inside the interval $[0, 6]$,

A MAXIMUM-MINIMUM PROBLEM

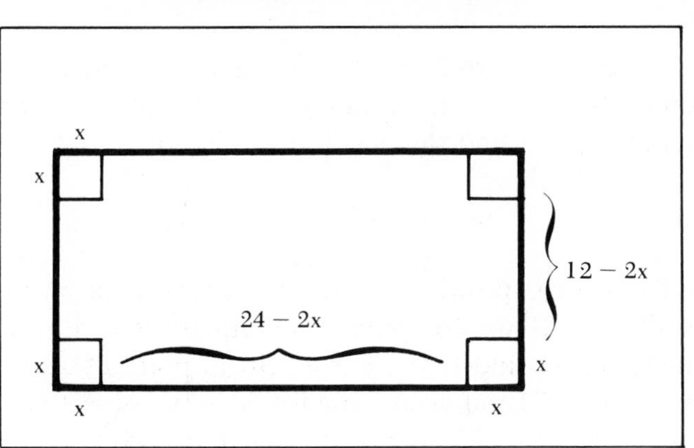

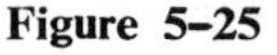

Figure 5–25

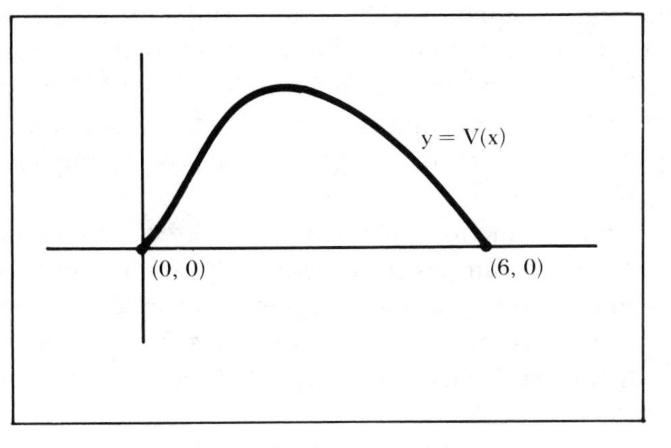

Figure 5–26

Figures 5–25 and **5–26** We are to find the volume of the largest box (without top) that can be made from a 12 inch by 24 inch sheet of cardboard. The box is constructed by cutting a square of side x inches from each corner of the sheet and then folding up the edges. We graph $V(x)$, the volume of the box obtained by letting the side of each square be x inches.

but it is a relative maximum of V. This can be verified by calculating the second derivative of V and noting that $V''(x) < 0$ for all $x < 6$. Therefore, the man should cut squares with side $6 - 2\sqrt{3}$ inches from the corners of his cardboard to maximize the volume of the resulting box.

Example 9: A manufacturer estimates that the profit $p(n)$ he will obtain from selling n items of a certain type is given by the equation

$$p(n) = -n^2 + 100n + 1.* \tag{16}$$

The manufacturer, of course, wants to know how many items he should sell to maximize his profits. Even though $p(n)$ is only defined when n is a positive integer, we will find the maximum value of

$$f(x) = -x^2 + 100x + 1 \tag{17}$$

and deduce from it the positive integer that is the solution to the manufacturer's problem. If we compute $f'(x)$, equate it to 0, and solve the resulting equation, we find that f has a critical point at $x = 50$. Since f is increasing for $x < 50$ and decreasing for $50 < x$, we see that $f(50)$ is the largest function value that f assumes. Thus f has a relative maximum at 50. Since 50 is itself an integer, we conclude that the manufacturer should sell 50 items to realize the maximum possible profit.

The function of equation (16) is an example of a *discrete* function. A discrete function is one that is defined for isolated real numbers in its domain: each point at which the function is defined has no other points close to it at which the function is also defined. A function defined only on a finite set, and a set of integers, would be examples of discrete functions.

Actually, most functions derived from direct observation of data are discrete, since we can generally take only finitely many readings of data. Often, once a discrete function has been obtained using observed data, we try to approximate it by a continuous function, as was done in Example 9. A further discussion of discrete functions appears in Section 6.4.

One of the basic properties of continuous functions—although it is one that we will not prove—is that if f is continuous on some closed interval $[a, b]$, then f will assume a largest function value and a smallest function value on $[a, b]$. Thus, there will be w and w' in $[a, b]$ such that $f(w') \leq f(x) \leq f(w)$ for every x in $[a, b]$. We should note that the maximum and minimum apply only to the interval $[a, b]$, for the function may have values *outside* $[a, b]$ larger or smaller than the largest and smallest values assumed on $[a, b]$.

* The function p is known as a *profit function*. We might ask why a profit function does not simply increase in direct proportion to the number of items sold, or in other words, why a profit function is not always of the form $p(n) = kn$, where k is the profit made on a single item. However, many factors can determine what profit or loss is realized with regard to some item. For example, when most of a magazine's revenue comes from advertising and the magazine is actually sold for less than the cost of production, then the magazine's publisher may actually wish to limit circulation. Yet adequate circulation must be maintained so that advertising rates can be kept suitably high. When the *Saturday Evening Post* suffered an increase in circulation together with a decline in advertising, it went bankrupt.

If we are asked to find the maximum or minimum value of some continuous function on a closed interval, then we can be sure that maximum and minimum function values exist for that closed interval. The problem is finding them. The following is a short program for solving this problem.

Suppose that f is a continuous function on the closed interval $[a, b]$, and we wish to find the relative maxima and minima of f on $[a, b]$.

(1) Find f'.
(2) Set $f'(x) = 0$ and solve this equation for all critical points of f. Discard those critical points not in $[a, b]$.
(3) Check the nature of each critical point of f in $[a, b]$ using the tests given in Proposition 3.
(4) Check a and b to determine whether f has a maximum or minimum at these points.
(5) Check any points of $[a, b]$ at which f is not differentiable to determine whether f has a relative maximum or minimum at any of these points.

Example 10: In Example 9 we found that if the manufacturer wants to maximize his profits, he should sell exactly 50 items. Suppose, however, that he had the capacity to produce at most 25 items. We would then have to restrict the variable x of equation (17) to the closed interval $[0, 25]$. Using the first derivative, we can easily show that $f(x) = -x^2 + 100x + 1$ is increasing on $[0, 25]$ (Figure 5–27). Hence, f will assume its maximum function value at the right endpoint of $[0, 25]$ (where $x = 25$). In this

A RELATIVE MAXIMUM AT THE ENDPOINT OF AN INTERVAL

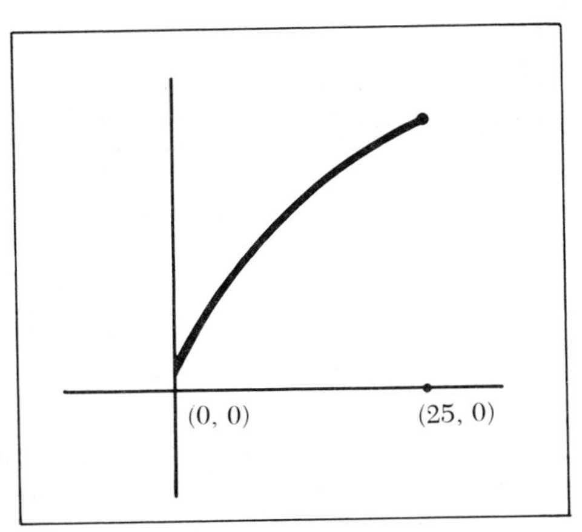

Figure 5–27 The graph of f on the interval $[0, 25]$ is rising; hence its maximum occurs at $x = 25$. Therefore, if the manufacturer can produce at most 25 items, he should produce 25 to realize maximum profits. The maximum does not occur at a critical point of the profit function.

case the manufacturer should produce the items at his full capacity in order to maximize his profits.

Example 11: Consider the absolute value function $f(x) = |x|$, where x is restricted to the closed interval $[-1, 2]$. The graph of f is shown in Figure 5–28. It is evident that the minimum function value of f occurs at $x = 0$ and the maximum function value occurs at 2, the right endpoint of the interval. The fact that $f(0) = 0$ is the minimum function value can be seen most readily by inspection of the definition of the function. Even though f has its minimum value at $x = 0$, this is not a critical point because f is not differentiable there. Using calculus, we can determine that f is decreasing on $(-1, 0)$ and increasing on $(0, 2)$; therefore, it seems reasonable to suppose that the maximum value of f will occur at one of the endpoints of the interval $[-1, 2]$. Again we see by inspection that f has its maximum value at $x = 2$.

A RELATIVE MINIMUM WHERE THE FIRST DERIVATIVE IS UNDEFINED

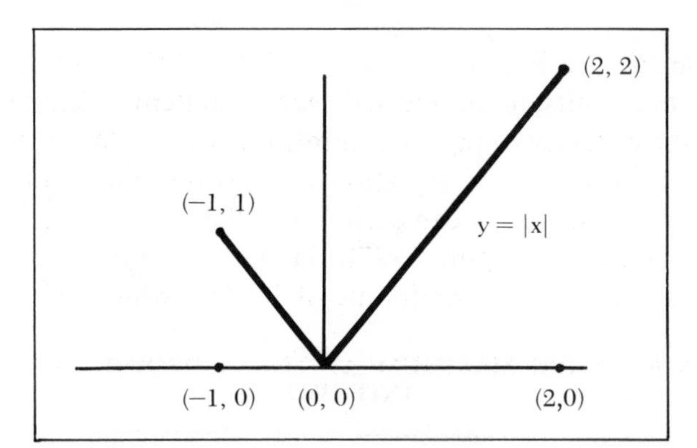

Figure 5–28 Note that $f(x) = |x|$ has a minimum at $x = 0$, but this minimum is not detectable using the calculus because f is not differentiable at $x = 0$. The minimum is easily found, however, by inspecting the graph of f.

Rates

Many functions encountered in practical applications of the calculus have time as their variable. For example, the following are all quantities that depend on time: the distance a moving object travels, the amount of money in an account earning a certain rate of interest, the number of emissions from a radioactive source, and the number of bacteria in a colony. The first derivative of a time-dependent function is itself a function of time. Since the first derivative in such cases generally measures how fast some quantity is changing with respect to time, the first derivative is sometimes called the *velocity* function. If we consider a distance function—a function that tells us how far an object has traveled after time t—the first derivative of the function is the actual velocity of the object at time t.

Example 12: A projectile is hurled straight upward with an initial velocity of 960 feet/second. Methods that will be outlined briefly later

A PROJECTILE IS HURLED STRAIGHT UPWARD WITH AN INITIAL VELOCITY OF 960 FEET/SECOND

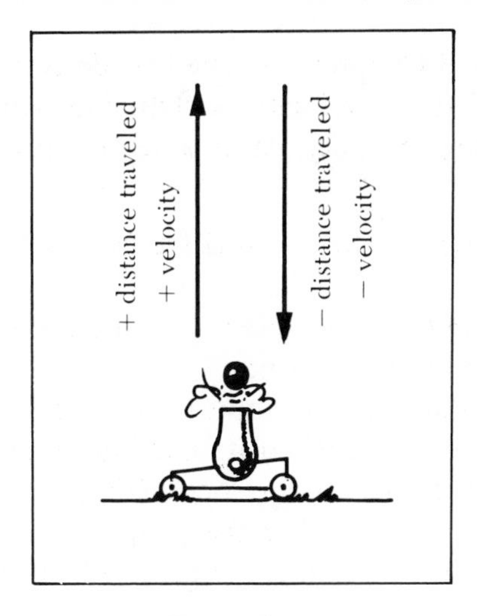

Figure 5–29

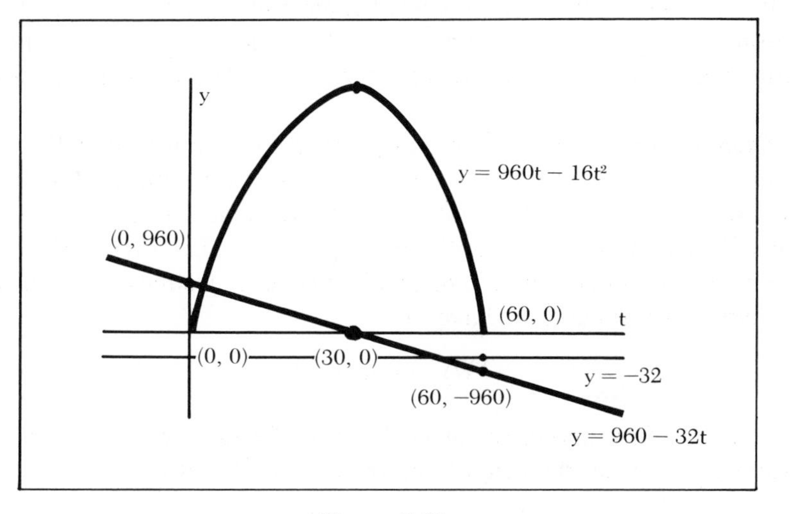

Figure 5–30

Figures 5–29 and **5–30** Upward distance and velocity are considered positive, while downward distance and velocity are negative. In Figure 5–30 the height of the projectile is plotted against the time elapsed since firing the projectile.

in this chapter enable us to conclude that t seconds after the projectile begins its flight, it will have traveled

$$s(t) = 960t - 16t^2 \text{ feet.} \tag{18}$$

The velocity of the projectile after time t is the first derivative of s, namely,

$$v(t) = s'(t) = 960 - 32t \text{ feet/second.} \tag{19}$$

(Review Example 33 of Chapter 1.) The first derivative of v, the function that indicates how fast the velocity is changing with respect to time, is the *acceleration* of the projectile. After time t the projectile is accelerating at rate $a(t)$, where

$$a(t) = v'(t) = -32 \text{ feet/second}^2. \tag{20}$$

According to equation (20), the velocity of the projectile is decreasing at the rate of 32 feet/second per second.

Note that the distance traveled upward is considered positive distance, but once the projectile has reached its maximum height, it begins losing distance and travels with negative velocity (Figure 5–29). The sign of the velocity function indicates whether the projectile is going up or coming down. In Figure 5–30, s, v, and a are graphed together on the same set of axes.

In certain instances, we may have several interrelated time-dependent functions. Often we can establish a relationship among their velocity functions and then use a known velocity to find an unknown velocity.

Example 13: Gas is being pumped into a spherical balloon at a rate of 4 cubic inches/minute. How fast is the radius of the balloon increasing when the balloon contains 20 cubic inches of gas? The volume of the balloon and its radius both depend on time. Let $V(t)$ be the volume of the balloon and $r(t)$ be its radius after time t. Using the equation for the volume of a sphere, we have

$$V(t) = \tfrac{4}{3}\pi(r(t))^3. \tag{21}$$

If we take the first derivative of both sides of equation (21), we obtain a relationship between the rate of change of V and that of r:

$$V'(t) = \tfrac{4}{3}\pi(3)(r(t))^2(r'(t)) = 4\pi(r(t))^2(r'(t)). \tag{22}$$

We are given $V'(t) = 4$ cubic inches/minute, and we are trying to find $r'(t)$ when $V(t) = 20$ cubic inches. By equation (21) we see that when

$V(t) = 20$, $r(t)$ is $(15/\pi)^{1/3}$. Substituting this value into equation (22), we obtain

$$4 = 4\pi(15/\pi)^{2/3}(r'(t)). \tag{23}$$

We can then solve equation (23) for $r'(t)$, which completes the solution.

Example 14: Two cars start from the same point at the same time. One car travels north at 30 miles per hour, while the other travels east at 50 miles per hour. At what rate are the cars drawing apart after 1 hour? (See Figure 5–31.)

THE CONDITIONS OF EXAMPLE 14

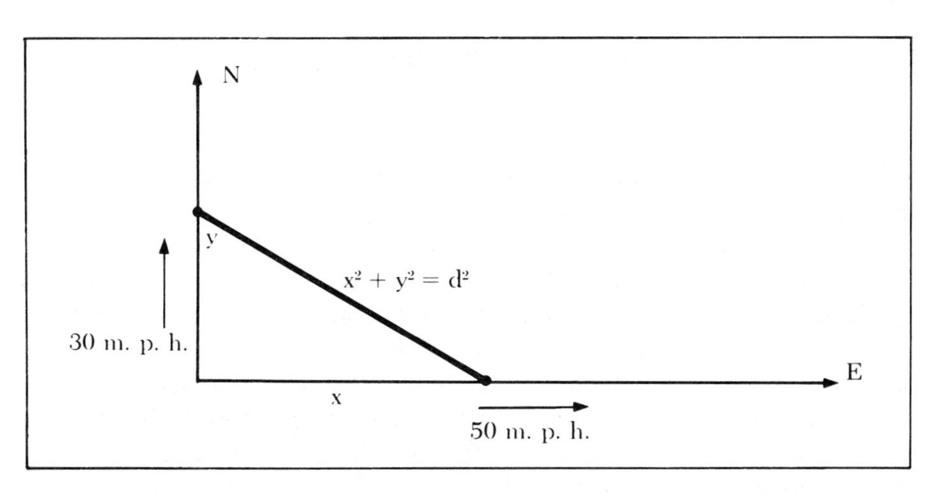

Figure 5–31 After time t, the northbound car has traveled $30t$ miles and the eastbound car has traveled $50t$ miles. Therefore, their distance from each other after time t is $((30t)^2 + (50t)^2)^{1/2}$ miles.

If the eastbound and northbound cars have gone x and y miles, respectively, then the Pythagorean Theorem can be used to find the distance d between the cars:

$$d^2 = x^2 + y^2 \tag{24}$$

Now x, y, and d are all dependent on time; hence equation (24) can be written as

$$(d(t))^2 = (x(t))^2 + (y(t))^2. \tag{25}$$

Furthermore, $d'(t)$ is the rate at which $d(t)$ is changing (the rate at which the cars are moving apart after time t). Thus $d'(1)$ is the solution to the problem. Computing the first derivative of both sides of equation (25), we see that

$$2d(t)\,d'(t) = 2x(t)x'(t) + 2y(t)y'(t). \tag{26}$$

The speed of the eastbound car is $x'(t) = 50$; that of the northbound car is $y'(t) = 30$. Moreover, the distance traveled after 1 hour by the eastbound car is $x(1) = 50$ miles, while that traveled by the northbound car is $y(1) = 30$ miles. Substituting these values into equation (25), we see that

$$\begin{aligned} d(1) &= ((x(1))^2 + (y(1))^2)^{1/2} \\ &= (2500 + 900)^{1/2} \\ &= (3400)^{1/2}. \end{aligned}$$

Substituting this into equation (26), we obtain

$$\begin{aligned} 2(3400)^{1/2}d'(1) &= 2(50)(50) + 2(30)(30) \\ &= 6800. \end{aligned} \tag{27}$$

Solving for $d'(1)$, we find that after 1 hour the cars are moving apart at the rate of $(3400)^{1/2}$ miles per hour.

Suppose we want to solve some problem involving the practical application of rates or maxima or minima, and we wish to apply the first derivative as discussed in this section. We must first translate the problem, which is usually stated verbally, into mathematical terms by finding a suitable variable and constructing an appropriate function in terms of that variable. Naturally, for the calculus to be applicable, the function in question must be differentiable. There is, unfortunately, no simple method of translating such problems into precise and usable mathematical forms. Generally, each problem must be individually analyzed. Recognizing techniques of the calculus that are appropriate for solving a given problem and then expressing the problem in terms that lead to a solution are skills that come from practice and some insight into the nature of the first derivative.

In the next section we will continue our discussion of applications of the first derivative.

Exercises

ROUTINE

1. Find two numbers whose sum is 20 and the sum of whose squares is the smallest possible number.

2. Find two numbers whose product is 20 and the square of whose sum is the smallest possible number.

3. A rock is thrown straight up with an initial velocity of 20 feet/second. After time t it has traveled $s(t) = 20t - 16t^2$ feet. How fast and in what direction is the rock moving after 1/2 second and after 2 seconds? How high does the rock go? How long is it in the air?

4. A man 6 feet tall walks away from a 10-foot lamppost at a uniform rate of 4 feet/second. How fast is the man's shadow lengthening when he is 10 feet from the lamppost? [Hint: Use similar triangles as suggested in Figure 5–32. You are looking for $L'(t)$.]

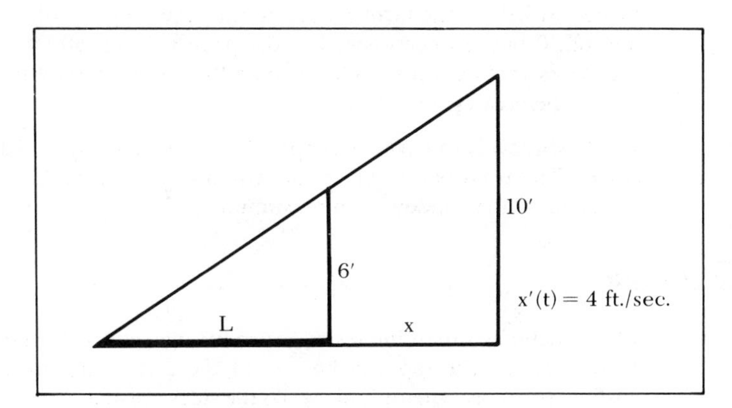

Figure 5–32 Exercise 4.

5. A wire 24 inches long is to be cut in two; one part is to be bent into a circle and the other into a square. How should the wire be cut if the sum of the areas of the circle and square is to be a minimum? a maximum?

6. Find the point on the graph of $y^2 = 4x$ that is nearest the point (2, 1).

7. Find the dimensions of the rectangle of largest perimeter that can be inscribed in a circle of radius 3.

8. A circular bar with radius 3 inches is to be machined to the shape of a rectangle. The manufacturer wants to have as little scrap material as possible. What should be the dimensions of the rectangle? [Hint: What are the dimensions of the rectangle of largest area that can be inscribed in a circle of radius 3?]

9. A manufacturer of portable radios estimates that his production costs for n radios are described by the equation $p(n) = 10n + 0.04n^2$. He also estimates that if he prices his radios to sell for $\$q$ each, he will sell $1000 - q/2$ radios. At what price should the manufacturer sell the radios to realize maximum profit? [Hint: At price $\$q$, the manufacturer will realize gross sales of $\$(1000 - q/2)q$. His cost of producing $1000 - q/2$ radios can be found by substituting $1000 - q/2$ for n in the equation for $p(n)$. The profit is the gross sales minus the production cost; this is the quantity to be maximized.]

10. A shoe manufacturer estimates that it costs $9n + 0.1n^2$ dollars to produce n pairs of shoes. If he prices a pair of shoes at $\$p$, he estimates that he will sell $900 - 0.05p$ pairs. At what price should he sell a pair of shoes in order to realize maximum profit?

11. A farmer wants to fence off 8000 square feet of his property. The enclosed area is to be in the shape of a rectangle, one side of which will run along the boundary of a neighboring farm. The neighboring farmer agrees to pay half the cost of the fence that runs along his boundary. For what dimensions of the rectangle will the first farmer have to pay the least?

12. A farmer has a crop of 500 bushels of potatoes that can be sold for \$2 per bushel. If the farmer waits to pick the crop, it will increase at the rate of 20 bushels per week, but the price will fall \$0.15 per bushel for each week that he waits. When should the farmer harvest the potatoes to realize the greatest revenues?

13. A window is to have the shape of a rectangle surmounted by a semicircle. The total perimeter of the window is to be 5 feet. Find the dimensions of the window with maximum area.

CHALLENGING

14. A manufacturer is interested in producing a carton having a volume of 1 cubic foot. The top and bottom of the carton are made of material costing \$0.10 per square foot, while the sides are made of material costing \$0.05 per square foot. If the bottom of the box is to be square, find the dimensions of the cheapest box that can be produced.

15. What is the length of the shortest ladder that will touch a wall standing 1 foot behind a fence 8 feet high?

16. Beer is flowing into a rectangular tank at the rate of 20 cubic feet per minute. If the tank measures 20 feet by 40 feet at its base, at what rate will the beer level in the tank rise?

17. A plane is cruising at 500 miles per hour at an altitude of 6 miles. It passes directly over an observation post on the ground. How fast is the plane moving away from the observation post 15 minutes after it has passed directly overhead? [Warning: Note that the speed of the plane is stated in miles per hour, while the problem is posed in terms of minutes. Be sure to express all distances in terms of the same unit and all times in terms of the same unit.]

18. A company estimates that producing n amplifiers of a certain type will cost $c(n) = n^2 - 50n + 50$ dollars. The amplifier is to sell for \$90; hence the revenue from the sale of n amplifiers will be $r(n) = 90n$ dollars. The profit function in this case is $p(n) = r(n) - c(n) = -n^2 + 140n - 50$. Explain why this is the correct profit function. How many amplifiers must the company sell to maximize its profit?

19. Telephone lines are to be strung between towns A and B, which are situated on opposite sides of a river 1 mile wide. Town A lies 3 miles downstream from a point directly across the river from town B (Figure 5–33). Stringing the line along the river bank would cost \$5000 per mile,

while special underwater cable costs \$9000 per mile to install. What is the most economical route the phone company can use in setting up the lines?

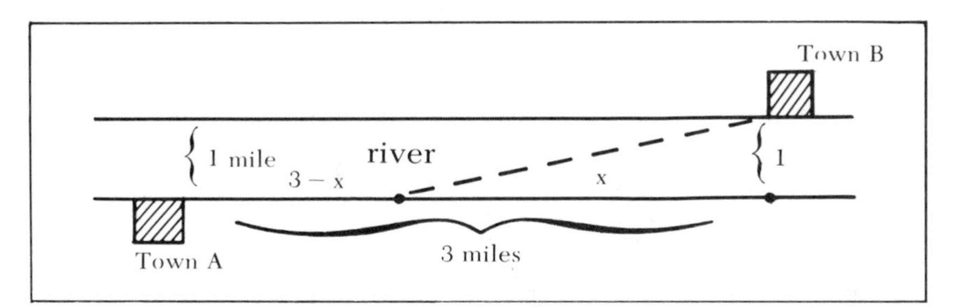

Figure 5–33 Exercise 20.

20. A Country Tax Commission wants to impose a sales tax of no less than 1 per cent but no more than 4 per cent on goods purchased in their county. The commission estimates that the average consumer who buys in their area spends about \$2000 per year on items that would be taxed, but if a tax rate of p per cent is imposed, the average will drop to \$2000/(1 + p). At what rate should the commission set the tax in order to realize maximum sales tax revenues?

5.4 FURTHER USES OF THE DERIVATIVE

Approximations

We have already seen that if a function f is differentiable at $x = a$, then the line with slope $f'(a)$ passing through the point $(a, f(a))$ is an approximation to the graph of f near $(a, f(a))$ (Figure 5–34). Since the equation of the line with slope $f'(a)$ passing through $(a, f(a))$ is

$$y = f'(a)(x - a) + f(a). \tag{28}$$

we can use equation (28) to estimate $f(x)$ for x close to a. The closer x is to a, the more accurate the estimate should be

We can also justify equation (28) as an approximation to $f(x)$ in the following manner: We know that the limit of

$$\frac{f(x) - f(a)}{x - a} \tag{29}$$

as x approaches a is $f'(a)$. Therefore, for x close to a, expression (29) should approximate $f'(a)$. The symbol $\doteq$ means *is approximately equal*

THE TANGENT LINE TO THE GRAPH OF f AS AN APPROXIMATION

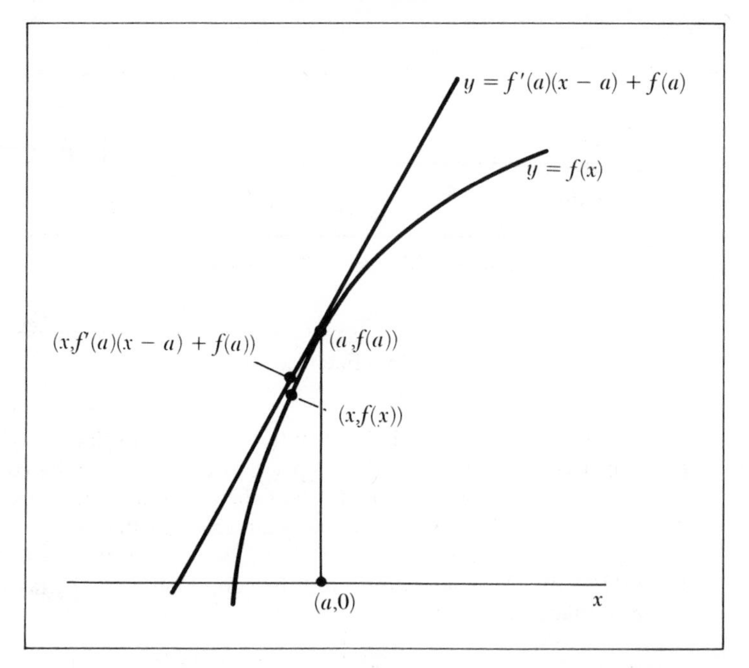

Figure 5–34 The tangent line to the graph of f at $(a, f(a))$ has the equation $y = f'(a)(x - a) + f(a)$. Since the tangent line is an approximation to the graph of f near $(a, f(a))$, we can use $f'(a)(x - a) + f(a)$ as an approximation to $f(x)$ for x close to a.

to; thus for x close to a,

$$\frac{f(x) - f(a)}{x - a} \doteq f'(a).$$

or

$$f(x) \doteq f'(a)(x - a) + f(a). \tag{30}$$

Why, however, would we want to approximate $f(x)$ if we can find $f(x)$ exactly? For example, there would be no need to use the above outlined methods to approximate $f(1.01)$ for $f(x) = x^2$ using $a = 1$ and $f'(1) = 2$ (and hence obtain $f(x) \doteq 2x - 1$ for x close to 1), since it is an easy matter to square 1.01 and find $f(1.01)$ exactly.

However, not all functions can be evaluated exactly or with the same facility as $f(x) = x^2$. While $(1.01)^2$ may be readily computable, $e^{1.01}$ and $(1.01)^{1/2}$ are not as easily found; moreover, since these latter numbers are irrational, any decimal representations of them will necessarily be approximations anyway.

Example 15: We now use the first derivative to approximate $e^{-0.01}$. First let $a = 0$. Since $f(x) = e^x$ and $f'(x) = e^x$, it follows that $f'(0) =$

$e^0 = 1$. Substituting these values into equation (30), we find that

$$f(x) = e^x \doteq (1)(x - 0) + 1 = x + 1.$$

When $x = -0.01$, which is reasonably close to 0, we see that $e^{-0.01} \doteq -0.01 + 1 = 0.99$. The actual value of $e^{-0.01}$ correct to five decimal places is 0.99005; hence our approximation is very good in this case.

Note that approximating a function by using the first derivative involves only simple arithmetic operations, assuming that $f(a)$ and $f'(a)$ are easily evaluated as in Example 15. Computers operate in such a way that the results they obtain must be obtainable by simple arithmetic operations. Thus, it is important to have techniques for simplifying such complex operations as finding roots and raising to a power.

It should be apparent, however, that the linear approximation discussed above is of limited value. Its effectiveness is severely limited by the fact that x must be close to a and $f(a)$ and $f'(a)$ must be readily evaluated in order to effect any real simplification of the computations. In hopes of getting a still better approximation that will hold for more values of x than will equation (30), we might try to approximate the graph of f near the point $(a, f(a))$ by something other than a straight line. In particular, we can look for a parabola that passes through $(a, f(a))$; to obtain the desired approximation, we require that the parabola have *the same first and second derivatives* at $x = a$ as the function being approximated.

We can show that

$$g(x) = \frac{f''(a)}{2}(x - a)^2 + f'(a)(x - a) + f(a) \tag{31}$$

PARABOLIC APPROXIMATION TO f

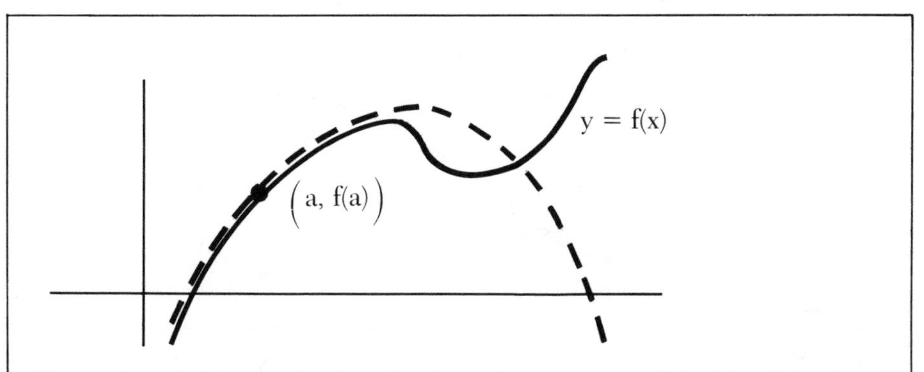

Figure 5–35 The parabola $y = [f''(a)/2](x - a)^2 + f'(a)(x - a) + f(a)$ approximates the graph of f near $(a, f(a))$. The function value, as well as the first and second derivatives of the parabola, coincide with those of f when $x = a$.

has the following properties: The graph of g is a parabola; $(a, f(a))$ is a point of the graph of g; $g'(a) = f'(a)$; and $g''(a) \doteq f''(a)$. (The graph of g is shown superimposed on the graph of f in Figure 5–35.) We would therefore expect $g(x)$ to approximate $f(x)$ for x close to a. Moreover, since g has the same second derivative as well as the same first derivative as f, while $f'(a)(x - a) + f(a)$ has only the same first derivative as f, we would expect the approximation given by equation (31) to be better than that given by equation (30). Also, equation (31) should provide good approximations for more values of x than does equation (30).

Example 16: We will approximate $(1.01)^{1/2}$ using equation (31). Let $f(x) = x^{1/2}$ and $a = 1$. Then $f(1) = 1, f'(1) = 1/2$, and $f''(1) = -1/4$. Substituting these values into equation (31), we see that

$$f(x) \doteq (-1/8)(x - 1)^2 + (1/2)(x - 1) + 1.$$

We therefore estimate $f(1.01)$ to be $(-1/8)(0.01)^2 + (1/2)(0.01) + 1 = 1.0049875$. The actual value of $(1.01)^{1/2}$ accurate to four decimal places is 1.0050; hence our approximation is quite accurate.

Although we will not pursue the matter further, we can use polynomials of successively higher degree to obtain increasingly better approximations of a function f. For any positive integer n, we could use a polynomial of degree n whose graph includes the point $(a, f(a))$ and has its first n derivatives at a equal to the first n derivatives of f at a. The sequence of approximating polynomials up to $n = 3$ is illustrated in Figure 5–36.

METHODS OF APPROXIMATING THE GRAPH OF f NEAR $(a, f(a))$

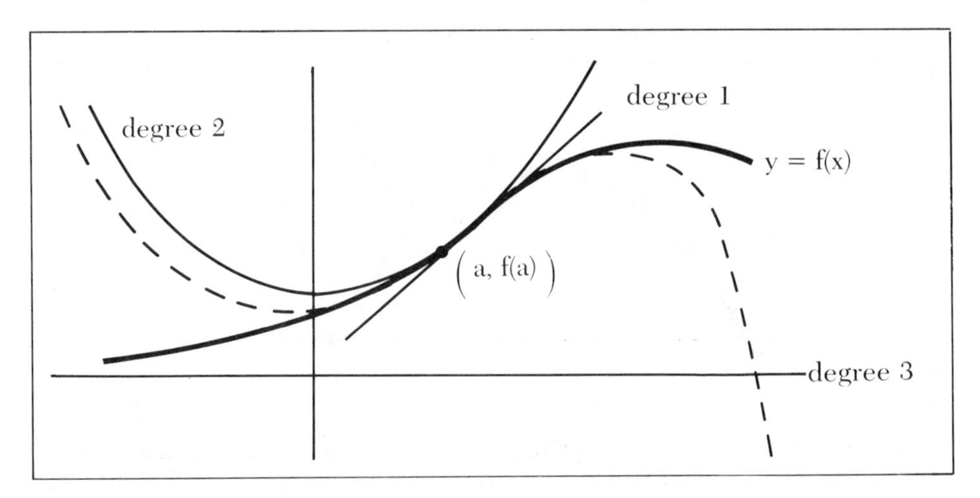

Figure 5–36 The graph of f near $(a, f(a))$ can be approximated by a straight line, a parabola, and the graphs of higher degree polynomials.

Differential Equations

In certain situations information is available concerning the first or higher derivatives of a function, and the function itself is to be found on the basis of this information. The next example illustrates such a situation.

Example 17: The population of a country and the number of microbes in a colony are examples of quantities whose rates of change are generally directly proportional to the quantity itself. Thus, if $P(t)$ is the population of the United States at time t, then $P'(t)$, the rate at which the population is changing at time t, is $kP(t)$, where k is some constant. The equation

$$P'(t) = kP(t) \tag{32}$$

is a specific mathematical way of saying that the more people there are, the faster the population increases.

Equation (32) is an example of a *differential equation;* it expresses a relationship between $P(t)$ and its first derivative, $P'(t)$. We will now try to solve equation (32) by finding a function P which satisfies it.

We know that $P(t) = e^{kt}$ is a function such that $P'(t) = ke^{kt} = kP(t)$; therefore

$$P(t) = e^{kt} \tag{33}$$

is a solution for equation (32). However, if we add any constant to kt in equation (33), we again obtain a solution for equation (32). The most general solution to (32) is

$$P(t) = e^{kt+c} \tag{34}$$

where c is a constant.

To evaluate c, we must know the value of $P(t)$ for some specific value of t. For example, if $P(0) = 100{,}000{,}000$, then equation (34) becomes

$$P(0) = 100{,}000{,}000 = e^{0+c} = e^c,$$

or

$$c = \log_e 100{,}000{,}000.$$

The condition $P(0) = 100{,}000{,}000$ is called an *initial condition.* The initial condition allows us to solve equation (32) completely. The complete solution to equation (32) with the given initial condition is

$$\begin{aligned} P(t) &= e^{kt+\log_e 100{,}000{,}000} \\ &= 100{,}000{,}000e^{kt}. \end{aligned}$$

If no initial condition is specified or presumed, then our solution must be left in the general form (34).

If one accepts equation (32) as true with regard to the population of the United States (and, hence, that the population of the United States may be computed at time t by an expression of the form given in (34)), then the constant k can be approximated experimentally to obtain a function for estimating the population of the United States at time t. We may also choose any point in time as the reference point $t = 0$. $P(0)$ is then the population of the United States at $t = 0$.

In Example 17 we had an equation involving a function and its first derivative. We then solved this equation for the function.

Definition 4: *An equation involving some function and its derivatives is called a* ***differential equation.*** *A function that satisfies a differential equation is commonly called an* ***integral*** *of the equation.*

The function of equation (33) is an integral of the differential equation (32). The most general integral for equation (32) is given in equation (34).

Example 18: It can be determined experimentally that a freely falling object accelerates at a rate of 32 feet/second². We can use this information to find the velocity of a falling object and the distance it has traveled at time t. Acceleration is the rate of change of velocity with respect to time; velocity is the rate of change of distance with respect to time. If $d(t)$, $v(t)$, and $a(t)$ designate distance, velocity, and acceleration at time t, then we can construct the following differential equations:

$$v'(t) = a(t) = 32 \qquad (35)$$

and

$$d'(t) = v(t). \qquad (36)$$

From equations (35) and (36) we obtain

$$d''(t) = a(t) = 32. \qquad (37)$$

The most general solution to equation (35) is

$$v(t) = 32t + c, \qquad (38)$$

where c is a constant. Substituting this general solution for $v(t)$ in equation

(36), we find that

$$d'(t) = 32t + c. \tag{39}$$

The most general solution to equation (39) and, hence, also to equation (37), is

$$d(t) = 16t^2 + ct + c', \tag{40}$$

where c and c' are constants.

In order to calculate c and c' in equations (38) and (40), we need initial conditions. To be able to compute c, we need the value of $v(t)$ for some specific value of t. Since two separate constants occur in equation (40), we need values of $d(t)$ for two distinct values of t to be able to compute them both. If we assume that the velocity and distance traveled are both 0 at time 0, then we have sufficient information to calculate c and c' exactly; under these assumptions both c and c' will be 0. Thus, if we assume $d(0) = 0$ and $v(0) = 0$, then

$$d(t) = 16t^2$$

becomes the integral for equation (37).

If we construct our frame of reference so that downward acceleration is negative (that is, -32 feet/second²) and incorporate $v(0) = 960$ feet/second instead of $v(0) = 0$, we can use equations (38) and (40) to derive equations (18) and (19) in Example 12 of this chapter.

Differential equations are generally more complex than those presented in Examples 17 and 18, and a variety of methods is available for solving them. However, a discussion of the means of solving differential equations requires a book in itself. Nevertheless, one special kind of differential equation will be of particular interest in the next chapter namely, the equation of the form

$$f'(x) = g(x). \tag{41}$$

To solve equation (41) we must find a function f whose derivative is the given function g.

Definition 5: *A solution f (when a solution exists) to equation (41) is called an* ***indefinite integral*** *of g and is often denoted by*

$$\int g(x)\, dx.$$

In the next chapter we will discuss methods of finding indefinite integrals and the reasons for their special importance.

Exercises

ROUTINE

1. We have seen (in Example 15) that the function $g(x) = x + 1$ approximates $f(x) = e^x$ for x close to 0.
   a) Find a polynomial of degree 2 that approximates e^x for x close to 0. See equation (31).
   b) Use the polynomial obtained in part (a) to approximate $e^{-0.01}$, and compare this approximation with the approximation to $e^{-0.01}$ found in Example 15.
   c) Use the polynomial obtained in part (a) to approximate $e^{0.1}$. Approximate $e^{0.1}$ using $g(x) = x + 1$. Look up $e^{0.1}$ in a table and determine which approximation is more accurate.
   d) Sketch the graph of the polynomial obtained in part (a) together with the graph of e^x on the same set of axes, and observe how the polynomial approximates e^x for x close to 0.

2. Find an indefinite integral for each of the following functions. In each case you will be looking for a function whose first derivative is the given function. Once you find such a function f, the general function will be f plus a constant term.
   a) $g(x) = 0$
   b) $g(x) = 1$
   c) $g(x) = x$ [Hint: Note that $f(x) = x^2$ has a first derivative twice as large as we want; hence we seek a function only half as large, namely $(1/2)f$.]
   d) $g(x) = x^2$
   e) $g(x) = x^2 - 3$
   f) $g(x) = 1/x$
   g) $g(x) = e^x$
   h) $g(x) = e^x + x$
   i) $g(x) = x^{1/5}$
   j) $g(x) = 34(x^2 + 1)^{33}(2x)$
   k) $g(x) = e^{x^3}(3x^2)$

CHALLENGING

3. Set up a differential equation characterizing each of the following situations. Solve the differential equation, if possible, and indicate any initial conditions stated implicitly or explicitly.
   a) The distance an object has traveled is inversely proportional to its speed.
   b) The distance an object has traveled is directly proportional to the square of its velocity.
   c) The time an object has traveled is directly proportional to its acceleration.
   d) The time an object has traveled is directly proportional to the sum of its velocity and acceleration. Moreover, its velocity and distance traveled at time 0 are both 0.

4. An object is thrown upward with an initial velocity of 40 feet per second. Because of gravity, the object accelerates at -32 feet/second per second. Note that the acceleration is negative because gravity acts in a direction opposite that in which the object was originally moving. Derive an equation showing the distance of the object from the ground after time t.

THEORETICAL

5. Suppose that f is a function for which $f^{(3)}(a)$ exists. Show that

$$g(x) = (f^{(3)}(a)/6)(x - a)^3 + (f''(a)/2)(x - a)^2 + f'(a)(x - a) + f(a)$$

has the following properties:

a) $f(a) = g(a)$; $f'(a) = g'(a)$; $f''(a) = g''(a)$; and $f^{(3)}(a) = g^{(3)}(a)$.

b) Compute $g(x)$ for $f(x) = e^x$ and $a = 0$.

c) Use the polynomial obtained in part (b) to approximate e^1. (The number e is approximately 2.71.)

6. Find a and b such that $f(x) = a(x - 4) + b$ can be used to approximate $\sqrt{4.1}$. Find A, B, and C such that $g(x) = A(x - 4)^2 + B(x - 4) + C$ can be used to approximate $\sqrt{4.1}$. Approximate $\sqrt{4.1}$ using both f and g. Which function gives the best approximation?

Review of Chapter 5

The first derivative f' of a function f measures the rate of change of f. If $f'(x) > 0$ for all x in (a, b), then f is *strictly increasing* on (a, b). If $f'(x) < 0$ for all x in (a, b), then f is *strictly decreasing* on (a, b).

A point a at which $f'(a) = 0$ is called a *critical point* of f. A differentiable function will have a critical point at b if f has either a relative maximum or minimum at b and if f is defined in some open interval that contains b. There are various tests for determining whether a function has either a relative maximum or minimum at any particular critical point.

The first derivative can be applied to graphing, problems involving rates, maximum-minimum problems, approximations, and differential equations.

In graphing a differentiable function it is helpful to know that the derivative indicates where the graph is rising, where it is falling, and where the maxima and minima of the function occur.

The first derivative of a function with time as the variable is called a *rate*, or *velocity*. Certain problems involving time and rates can be solved by the use of derivatives, providing one can find a suitable equation to describe the problem in terms of differentiable functions.

Use of the derivative for approximations is based on the fact that, when x is close to a, $f'(a)(x - a)$ is approximately equal to $f(x) - f(a)$.

Sometimes information about the derivatives of a function can be expressed in an equation; such an equation is called a *differential equation.* The function itself can then be found by solving the differential equation.

REVIEW EXERCISES

1. Graph each of the following functions. Indicate (i) where critical points occur and the nature of each critical point; (ii) where the function is increasing and where it is decreasing; and (iii) any points at which the function is not differentiable.

a) $f(x) = 3x - 9$
b) $f(x) = 3e^x - 9$
c) $f(x) = 3 \log_e x - 9$
d) $g(t) = (\log_e x)^2$
e) $f(x) = 4x,\ 3 \leq x \leq 7$
f) $f(x) = 4x,\ 3 < x < 7$
g) $g(t) = e^t - t^2$
h) $h(t) = t/(1 + t^2)$

2. A boat is sailing due north at 20 m.p.h., while a car is traveling due east at 50 m.p.h. How fast are the car and boat separating when they are exactly 10 miles apart?

3. The surface area of a spherical dome of radius r is given by $S = 2\pi r^2$. An engineer estimates the radius of a certain spherical dome to be 100 feet,

whereas the radius is actually 101 feet. Use S' to estimate the difference between the engineer's estimate of the surface area of the dome and the actual surface area.

4. Use the first derivative to find the value of n for which the maximum profit is realized in Exercise 3 of the review exercises for Chapter 1.

5. Express the following information by means of a suitable differential equation: The rate at which the profits of a certain corporation are increasing is inversely proportional to time. At the beginning of the current year, the corporation's profits were \$500,000. (Note that y is *inversely proportional* to x if $y = k/x$, where k is a constant.)

Vocabulary

definite integral
area
Fundamental Theorem of Calculus
substitution of variable
integration by parts

Chapter 6

The Definite Integral

6.1 THE CONCEPT OF THE DEFINITE INTEGRAL

Example 1: Suppose that a vehicle is moving at time t with velocity

$$v(t) = t. \tag{1}$$

If $d(t)$ denotes the distance the vehicle has traveled at time t, then $d(t)$ is the solution to the differential equation

$$d'(t) = v(t) = t.$$

We conclude that

$$d(t) = t^2/2 + c, \tag{2}$$

where c is a constant. We cannot evaluate c without assuming or being given a value of $d(t)$ for a specific value of t; but we can, although c is unknown, answer the question: "How far does the vehicle travel between time $t = 0$ and $t = 1$?" The total distance traveled between $t = 0$ and $t = 1$ is simply

$$\begin{aligned} d(1) - d(0) &= (1^2/2 + c) - (0^2/2 + c) \\ &= (1/2 - 0) + (c - c) \\ &= 1/2. \end{aligned}$$

The constant term c cancels out in the computation. We now employ a different technique to find the distance the vehicle travels between $t = 0$ and $t = 1$.

If a vehicle travels with constant velocity k, then in time t the vehicle travels a total distance kt. For example, an automobile traveling at 30 miles/hour will cover $30t$ miles in t hours. An automobile whose velocity is 0 will travel 0 distance between $t = 0$ and $t = 1$. On the other hand, an automobile with velocity 1 will travel a total distance of $1 \cdot 1$ units from $t = 0$ to $t = 1$.

A vehicle whose velocity is given by equation (1) has velocity 0 at $t = 0$. By the time $t = 1$, the velocity has increased to 1. We would expect the total distance traveled to be somewhere between 0 and 1, and such is the case.

Although the velocity of the vehicle is not constant from $t = 0$ to $t = 1$, we can subdivide the closed interval [0, 1] into subintervals small enough so that the velocity can be considered approximately constant over each subinterval. If $[t_{i-1}, t_i]$ is the ith such subinterval, we will let the velocity be $v(t_i)$ over this entire subinterval. Therefore, the total distance the vehicle travels from t_{i-1} to t_i, assuming a constant velocity $v(t_i)$, is

$$v(t_i)(t_i - t_{i-1}) = \text{velocity} \cdot \text{time}. \tag{3}$$

Thus, if [0, 1] is divided as in Figure 6–1, then the total distance the

ESTIMATION OF VELOCITY OVER A SUBINTERVAL OF [0, 1]

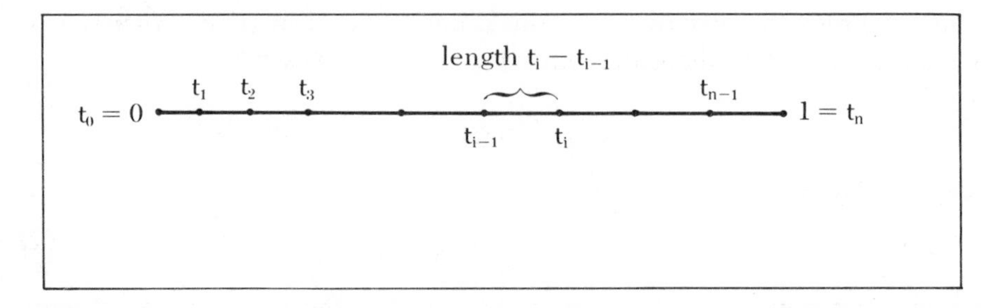

Figure 6–1 We divide the interval [0, 1] into n subintervals. The velocity over the ith subinterval $[t_{i-1}, t_i]$ is estimated by $v(t_i)$.

vehicle travels from $t = 0$ to $t = 1$ is approximated by the sum of the terms

$$v(t_i)(t_i - t_{i-1}) \quad \text{for} \quad i = 1, 2, 3, \ldots, n,$$

or

$$v(t_1)(t_1 - t_0) + v(t_2)(t_2 - t_1) + \ldots + v(t_n)(t_n - t_{n-1})$$
$$= t_1(t_1 - t_0) + t_2(t_2 - t_1) + \ldots + t_n(t_n - t_{n-1}), \tag{4}$$

since $v(t) = t$ for any t in [0, 1].

If the subintervals of [0, 1] are chosen sufficiently small, then equation (4) should yield a fairly accurate approximation to the total distance the vehicle travels from $t = 0$ to $t = 1$. We know already that this distance is 1/2; hence the smaller and more numerous the subintervals, the closer equation (4) should be to 1/2. Indeed, equation (4) approaches a limit of 1/2 as we divide [0, 1] into smaller and smaller subintervals.

Suppose we now divide [0, 1] into n *equal* subintervals (Figure 6–2).

A DIVISION INTO EQUAL SUBINTERVALS

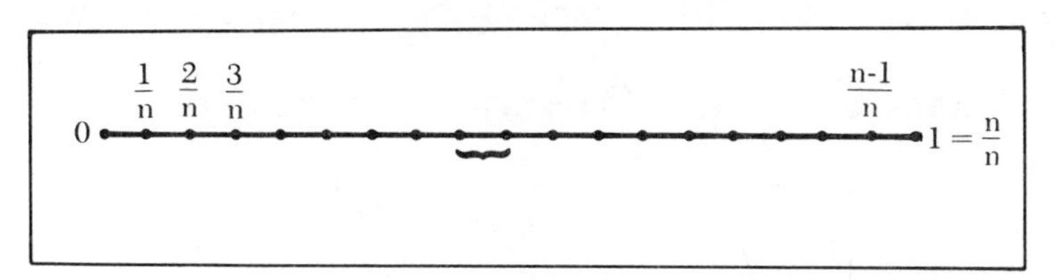

Figure 6–2 We now divide [0, 1] into n equal subintervals. Each subinterval will have length $1/n$.

Then each subinterval will have length $1/n$, and the endpoints will be

$$0, 1/n, 2/n, 3/n, \ldots, (n-1)/n, n/n = 1.$$

In this particular case the sum in equation (4) becomes

$$(1/n)(1/n) + (2/n)(1/n) + (3/n)(1/n) + \ldots + \left(\frac{n-1}{n}\right)(1/n) + (n/n)(1/n) = (1/n)^2(1 + 2 + 3 + \ldots + n).$$

Let $S(n)$ denote $1 + 2 + 3 + \ldots + n$. Then the last expression in equation (5) is $(1/n)^2S(n)$. Since

$$\begin{aligned} S(n) &= 1 + 2 + 3 + \ldots + n \\ +S(n) &= n + (n-1) + (n-2) + \ldots + 1 \\ \hline 2S(n) &= (n+1) + (n+1) + \ldots + (n+1)(n \text{ times}) = n(n+1), \end{aligned}$$

we conclude that $S(n)$ is $(n/2)(n + 1)$. Substituting for $S(n)$ in equation (5), we obtain

$$(1/n)^2S(n) = (1/n)(1/n)(n/2)(n+1) = (1/2)((n+1)/n). \qquad (6)$$

As n takes on increasingly large integer values, the subdivision of [0, 1] becomes finer. The limit of $(n + 1)/n$ is 1; hence the limit of equation (6) is 1/2. Thus, in this particular case, the approximations to the total distance traveled from $t = 0$ to $t = 1$ tend to 1/2 (the exact distance traveled) as we subdivide [0, 1] more and more finely.

We already were able to answer the question about the total distance traveled even before we resorted to the approximation and limit process above. Now we discuss a problem for which our previous work with derivatives offers no techniques of solution. However, a limit of approximations method such as that used above will lead to a solution.

Example 2: Consider the area of the plane bounded by the lines $x = 0$, $x = 1$, $y = 0$, and the graph of $y = x^2$ (Figure 6–3). We want to

AREA UNDER THE GRAPH OF A FUNCTION ON [0, 1]

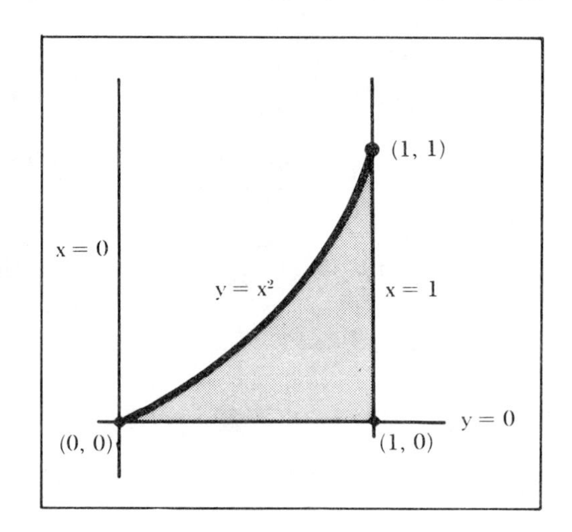

Figure 6–3 We wish to find a numerical value for the area bounded by the lines $x = 0$, $x = 1$, $y = 0$, and the graph of $y = x^2$.

assign a numerical value to this area. Since the area in question is not a regular geometric figure such as a triangle or rectangle, none of the usual formulas for computing area applies. Because the area in question lies inside a square of area 1 (Figure 6–4), we expect the area to be some number between 0 and 1. We will try to find the exact area by using an approximation–limit technique similar to that used in Example 1.

Actually, no definition of area has been given previously for a region such as that pictured in Figure 6–3. Strictly speaking, although we may have an intuitive notion of area, we are actually *defining* the concept of area for this region.

Again we divide the interval [0, 1] into subintervals (Figure 6–5). The portion of the area over the subinterval $[x_{i-1}, x_i]$ is approximated by a rectangle of height

$$f(x_i) = x_i^2.$$

To approximate the total area, we add the areas of all such rectangles. Since the rectangle over the interval $[x_{i-1}, x_i]$ has base $x_i - x_{i-1}$ and

A GEOMETRICAL ESTIMATION OF AREA

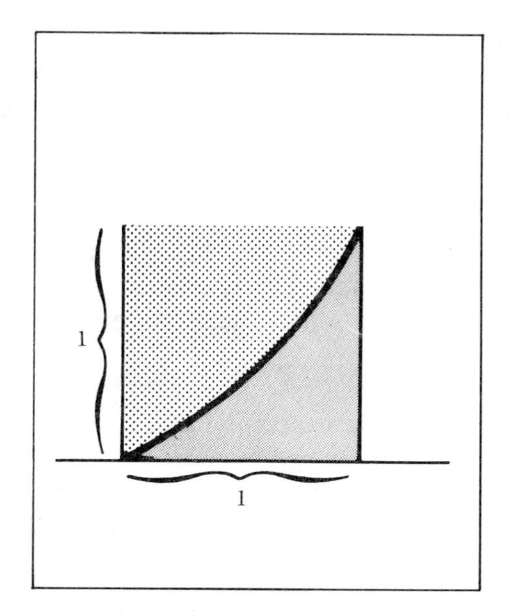

Figure 6–4 The area in Figure 6–3 is greater than 0; since it lies within a square of area 1, it must be less than 1.

APPROXIMATION BY RECTANGLES

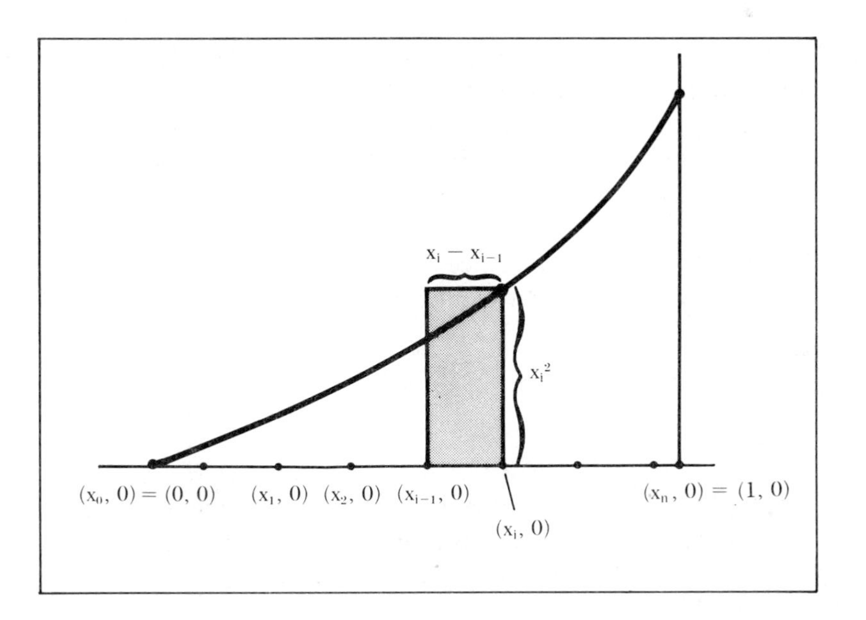

Figure 6–5 The portion of the area between $x = x_{i-1}$ and $x = x_i$ is approximated by the rectangle with width $x_i - x_{i-1}$ and height x_i^2. To approximate the entire area, we sum the areas of all such rectangles.

height x_i^2, the area is

$$x_i^2(x_i - x_{i-1}).$$

The approximation for the total area will be

$$x_1^2(x_1 - x_0) + x_2^2(x_2 - x_1) + x_3^2(x_3 - x_2) + \ldots + x_n^2(x_n - x_{n-1}). \quad (7)$$

This sum, of course, represents a number which depends on n. As we subdivide [0, 1] into finer and finer parts, the sum (7) should yield a better and better approximation of the area; if (7) has a limit, then this limit should be the exact area.

Again consider the special case in which [0, 1] is divided into n equal subintervals. In this instance expression (7) becomes

$$(1/n)^2(1/n) + (2/n)^2(1/n) + \ldots + ((n-1)/n)^2(1/n) + (n/n)^2(1/n)$$
$$= (1/n)^3(1^2 + 2^2 + 3^2 + \ldots + n^2). \quad (8)$$

The sum $1^2 + 2^2 + 3^2 + \ldots + n^2$ is given by $(1/6)n(n+1)(2n+1)$, the proof of which we omit. Therefore, the last expression in equation (8) is equal to

$$(1/n)(1/n)(1/n)(1/6)(n)(n+1)(2n+1) = (1/6)((n+1)/n)((2n+1)/n). \quad (9)$$

The limit of $(n+1)/n$ is 1 and the limit of $(2n+1)/n$ is 2. Therefore, equation (9) approaches the limit $(1/6)(2) = 1/3$ as we subdivide [0, 1] into more and more equal subintervals. The approximation for various values of n are illustrated geometrically in Figures 6–6 to 6–9. It is reasonable, then, to suppose that 1/3 is the actual area pictured in Figure 6–3.

In Example 1 the total distance traveled was $d(1) - d(0)$, where d was the indefinite integral of v. In Example 2,

$$F(x) = x^3/3 + c$$

is an indefinite integral of $f(x) = x^2$ for each value of the constant c. We can easily check this assertion by differentiating F. We note here that $F(1) - F(0) = 1/3$, the area in question and the solution to the problem. We might now conjecture that indefinite integrals can be used to compute not only distances, but areas, and perhaps many other unknowns as well.

The limits of the sums in equations (4) and (7) are particular instances of *definite integrals*. We now formulate a general definition for the definite integral.

THE PROCESS OF APPROXIMATION OF AREA

Figure 6–6 **Figure 6–7**

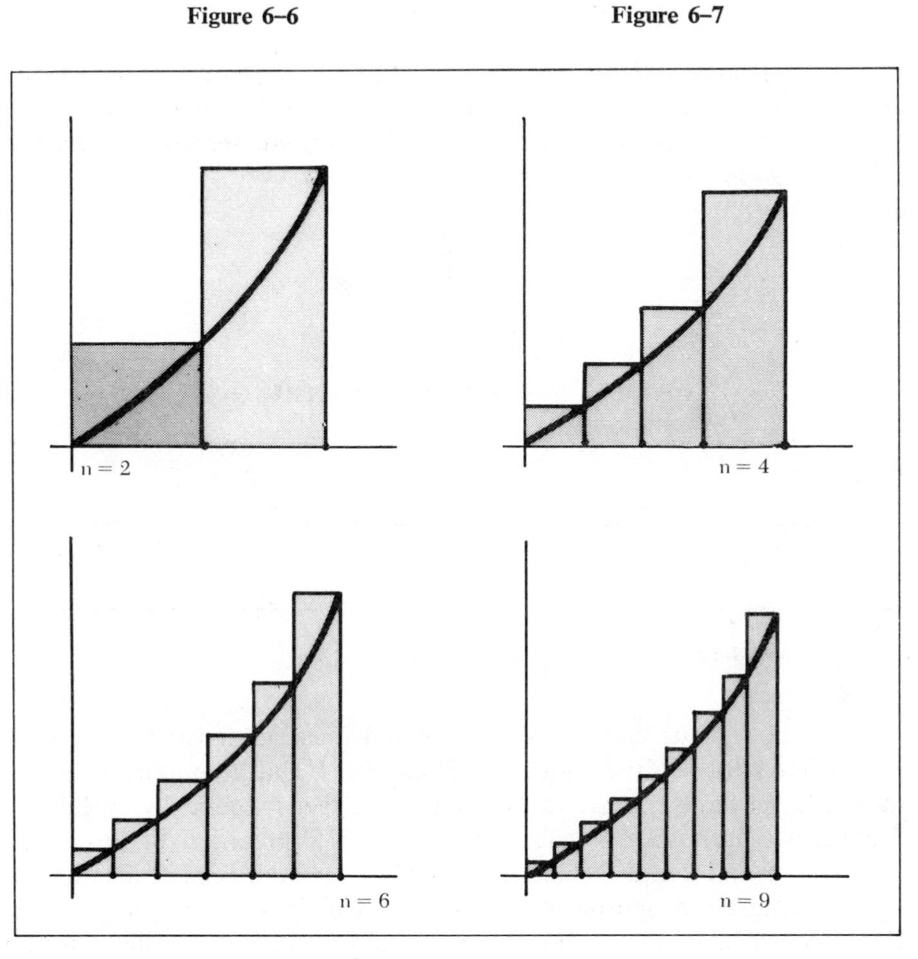

Figure 6–8 **Figure 6–9**

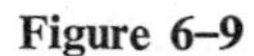

Figures 6–6, 6–7, 6–8, and **6–9** The approximation indicated in Figure 6–5 is easiest to compute if we divide the interval [0, 1] into n equal subintervals. The approximations for various values of n are shown. Note that the larger the value of n, the better the sum of the areas of the rectangles approximates the area of Figure 6–3.

Definition 1: *Let f be a function defined at each point of the closed interval $[a, b]$. For each subdivision of $[a, b]$ into subintervals (see Figure 6–10), we form the sum*

$$f(x_1)(x_1 - x_0) + f(x_2)(x_2 - x_1) + \ldots + f(x_n)(x_n - x_{n-1}). \quad (10)$$

As finer and finer subdivisions of $[a, b]$ are chosen (as the lengths of the individual subintervals approach 0), *the limit of expression* (10), *if the limit exists, is called the* ***definite integral of f from a to b.*** *It is denoted by*

$$\int_a^b f(x)\,dx.$$

DIVIDING [*a*, *b*] INTO *n* SUBINTERVALS

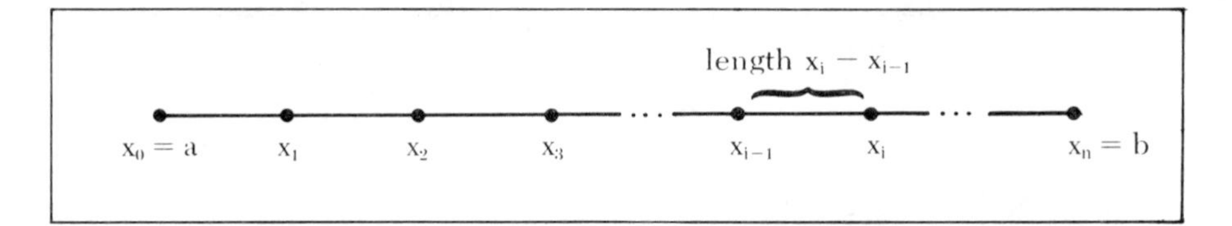

Figure 6–10 The length of the ith subinterval $[x_{i-1}, x_i]$ is $x_i - x_{i-1}$.

The notions of "limit" and "finer and finer subdivisions" are purposely left informal and intuitive. Examples 1 and 2 contain instances in which the limit appears to exist and is in fact numerically evaluated. Figures 6–6 through 6–9 offer a geometric interpretation of this limit. Examples 1 and 2 and Definition 1 should provide a general idea of the definite integral; a rigorous discussion of the definite integral lies beyond the intended scope of this book. We should also note here that the use of the endpoints $x_1, x_2, \ldots, x_n$ of the subintervals for evaluation of f in expression (10) is a very special choice, and any other choice of points in the subintervals would be permissible and would lead to exactly the same value of the definite integral (provided the definite integral exists) that expression (10) does.

We now turn our attention to the following questions, which flow naturally from Definition 1:

I) Under what conditions does the definite integral exist? What properties, in other words, must f possess in order for $\int_a^b f(x)\,dx$ to have a value?

II) How do we actually find $\int_a^b f(x)\,dx$ once we are sure that it exists?
III) What are some practical uses of the definite integral?

The next two sections will be concerned with answers to the first two questions. The last section of this chapter will give a partial answer to Question III.

The following example indicates the close relationship between the indefinite integral and the definite integral.

Example 3: If $f(x) \geq 0$ for all x in $[a, b]$, then expression (10) can be interpreted geometrically as an *approximation* to the area bounded by the graph of f, the x-axis, and the vertical lines $x = a$ and $x = b$ (Figure 6–11). Therefore, we are justified in interpreting $\int_a^b f(x)\,dx$ as the actual area within the boundaries indicated.

APPROXIMATION BY A SUM OF RECTANGULAR AREAS

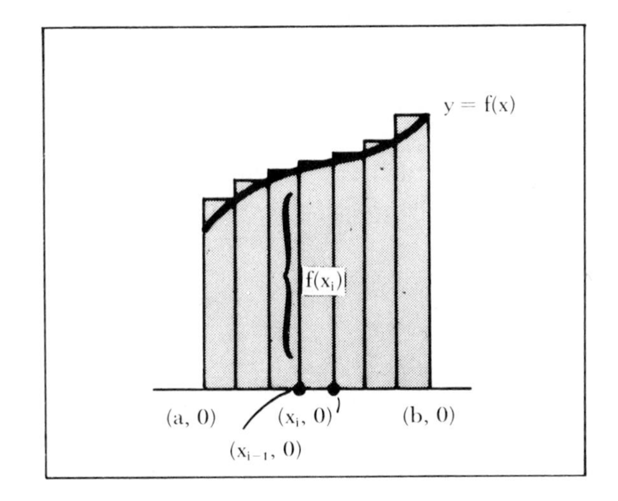

Figure 6–11 Geometrically interpreted, expression (10) is the sum of the areas of the rectangles shown in Figure 6–10. This sum approximates the area bounded by the lines $x = a$, $x = b$, $y = 0$, and the graph of $y = f(x)$. The limit of expression (10) as the subdivisions are chosen finer and finer is the area itself, and is also $\int_a^b f(x)\,dx$.

Suppose that f is defined on $[0, b]$ with $0 \leq a < b$, and let $A(w)$ denote the area bounded by the graph of f, the x-axis, and the line $x = 0$ and $x = w$ (Figure 6–12). Then A is a function of w (note that

$$A(w) = \int_0^w f(x)\,dx$$

for each w in $[a, b]$); we will try to find $A'(w)$.

THE INTEGRAL INTERPRETED AS AN AREA

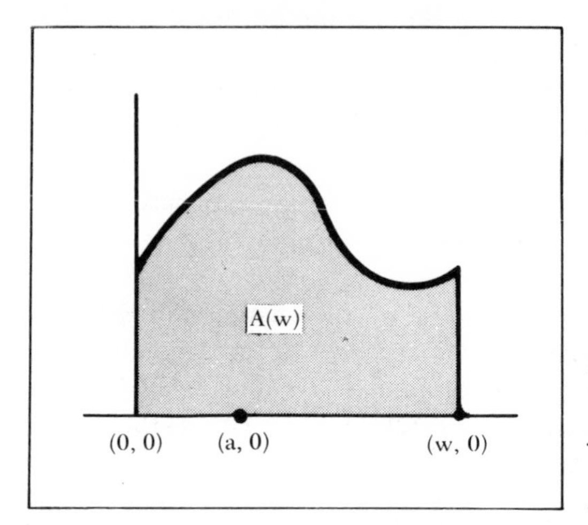

Figure 6–12 $A(w)$ denotes the area bounded by the graph of f, the y-axis, and the lines $x = 0$ and $x = w$. For each w in $[a, b]$, $A(w) = \int_0^w f(x)\,dx$.

Now

$$A(w + h) - A(w)$$

is approximated by the area of the rectangle of height $f(w + h)$ and base h (Figure 6–13), and the smaller h is, the better the approximation will be. The area of the rectangle is $f(w + h)h$. Therefore,

$$\frac{A(w + h) - A(w)}{h} \doteq f(w + h)h/h = f(w + h). \qquad (11)$$

USING AN APPROXIMATION TO FIND $A'(w)$

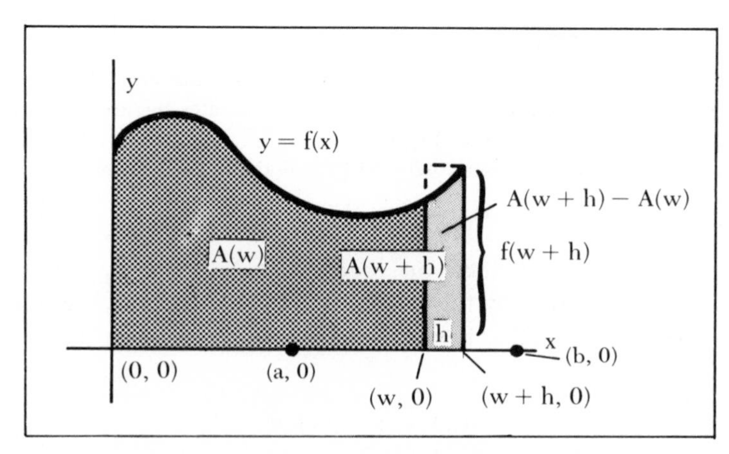

Figure 6–13 $A(w + h) - A(w)$ is approximated by the area of the rectangle of height $f(w + h)$ and base h. The smaller h is, the better the approximation.

The limit of the right side of equation (11) as h approaches 0 is $f(w)$, provided that f is continuous at w. Moreover, as h approaches 0, the left and right sides of equation (11) both approach the same limit $f(w)$. But the left side of (11) has $A'(w)$ as its limit. We therefore conclude by this intuitive argument that if f is continuous, then

$$A'(w) = f(w). \tag{12}$$

The area function A is actually the indefinite integral of f.

In terms of A, the area bounded by the graph of f, the x-axis, and the vertical lines $x = a$ and $x = b$ is simply $A(b) - A(a)$ (Figure 6–14); we

$\int_a^b f(x)\,dx = A(b) - A(a)$

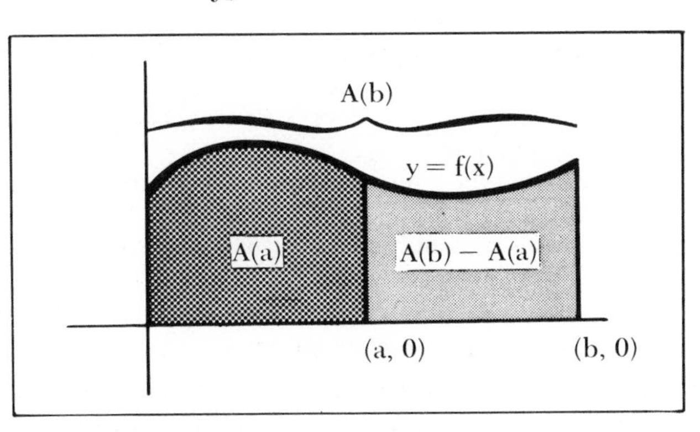

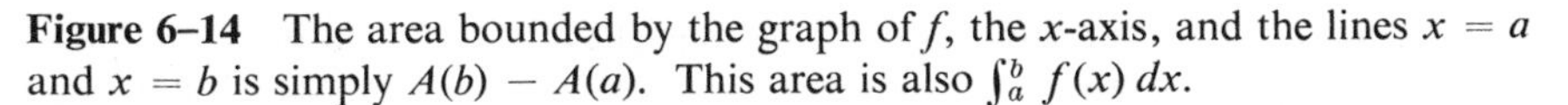

Figure 6–14 The area bounded by the graph of f, the x-axis, and the lines $x = a$ and $x = b$ is simply $A(b) - A(a)$. This area is also $\int_a^b f(x)\,dx$.

also know that this area is $\int_a^b f(x)\,dx$. We can therefore say that if f is continuous, then

$$\int_a^b f(x)\,dx = A(b) - A(a). \tag{13}$$

where A is the indefinite integral of f. In the next section we will see that equation (13) remains true if f is continuous on $[a, b]$ even if $f(x)$ is negative for some or all x in $[a, b]$.

Exercises

ROUTINE

1. Let $f(x) = x^2$. The endpoints of the subintervals in a subdivision of $[0, 2]$ are indicated in each of the following. In each case evaluate expression (10) for f and the specific subdivision.

a) 0, 1, 2
b) 0, 1/2, 1, 2
c) 0, 1/2, 1, 3/2, 2
d) 0, 0.2, 0.4, 0.6, 0.8, 1, 1.2, 1.4, 1.6, 1.8, 2
e) 0, 1, 3/2, 7/4, 15/8, 2

2. Sketch the area represented by the sum obtained in each part of Exercise 1. For examples of this type of sketch, see Figures 6–6 to 6–9.

3. Find the indefinite integral of $f(x) = x^2$. Use this indefinite integral to find the exact area bounded by the graph of f, the x-axis, and the lines $x = 0$ and $x = 2$ (cf. Example 3). Compare the exact area with the approximations obtained using the various subdivisions of [0, 2] given in Exercise 1.

4. a) Find the endpoints for a subdivision of [1, 2] that contains n equal subintervals.
b) Let $f(x) = x$ and write out expression (10) using the subdivision of [1, 2] obtained in (a).
c) What numerical limit does expression (10) approach as n increases? This limit is $\int_1^2 x\,dx$.
d) Represent the limit from part (c) geometrically. What area of the plane does it represent? Evaluate this area using the standard formulas of plane geometry; of course, the two computations of the areas—from parts (c) and (d)—should agree.
e) Find an indefinite integral of f and use this indefinite integral as in Example 3 to evaluate the area.

5. Repeat Exercise 4, replacing $f(x) = x$ by $g(x) = 2x + 1$ and the interval [1, 2] by [0, 4]. The integral to be found is now $\int_0^4 (2x + 1)\,dx$.

6. For each function given below, find the area bounded by the graph of the function, the x-axis, and the lines $x = 1$ and $x = 5$. Sketch the graph in each case.

a) $f(x) = x$
b) $f(x) = x^2$
c) $f(x) = x^3$
d) $f(x) = 2x$
e) $f(x) = 3x$
f) $f(x) = 2x + 1$
g) $f(x) = 2x + 2$
h) $f(x) = 2x + 3$

6.2 THE FUNDAMENTAL THEOREM OF THE CALCULUS

We now direct our attention toward determining when the definite integral exists and evaluating it when we know it exists.

Suppose that f is a continuous function on the closed interval $[a, b]$, and further that f has an indefinite integral F. There exists, given these conditions, a function F such that $F'(x) = f(x)$ for x in $[a, b]$. Assume that $[x_{i-1}, x_i]$ is a subinterval in some subdivision of $[a, b]$. Since F is differentiable on $[a, b]$, it is also continuous on $[a, b]$; consequently, we can apply the Mean Value Theorem (Proposition 4 of Chapter 5) to F on the interval $[x_{i-1}, x_i]$. In particular, we can say that there is some point

t_i in (x_{i-1}, x_i) such that

$$\frac{F(x_i) - F(x_{i-1})}{x_i - x_{i-1}} = F'(t_i) = f(t_i).$$

or

$$F(x_i) - F(x_{i-1}) = f(t_i)(x_i - x_{i-1}). \tag{14}$$

Now the right side of equation (14) closely resembles the expression $f(x_i)(x_i - x_{i-1})$ in the definition of the definite integral, expression (10). Indeed, because of the continuity of f, the sum

$$f(t_1)(x_1 - x_0) + f(t_2)(x_2 - x_1) + \ldots + f(t_n)(x_n - x_{n-1}) \tag{15}$$

will approach exactly the same limit as does

$$f(x_1)(x_1 - x_0) + f(x_2)(x_2 - x_1) + \ldots + f(x_n)(x_n - x_{n-1}) \tag{10}$$

as finer and finer subdivisions are chosen. Since the limit of (10) is by definition $\int_b^a f(x)\,dx$, the limit of (15) will also be $\int_a^b f(x)\,dx$.

The most we can do here, given our objectives of brevity and simplicity, is suggest that it is reasonable to assume that expressions (10) and (15) have the same limit—provided a limit exists—and that (15) can be used as readily as (10) to find $\int_a^b f(x)\,dx$. Observe that if f is continuous and the subinterval $[x_{i-1}, x_i]$ is fairly small, then *all* of the function values for x in $[x_{i-1}, x_i]$ will be approximately equal. Thus f will be approximately constant on $[x_{i-1}, x_i]$, so our assumption is reasonable.

Using equation (14), we determine that the sum in (15) is equal to

$$(F(x_1) - F(x_0)) + (F(x_2) - F(x_1)) + \ldots + (F(x_{n-1}) - F(x_{n-2})) \\ + (F(x_n) - F(x_{n-1})). \tag{16}$$

Regrouping the terms of expression (16), we obtain

$$-F(x_0) + (F(x_1) - F(x_1)) + (F(x_2) - F(x_2)) \\ + \ldots + (F(x_{n-1}) - F(x_{n-1})) + F(x_n). \tag{17}$$

Since x_0, the first end point in the subdivision of $[a, b]$, is always a, while x_n, the last end point, is always b, expression (17) reduces to just

$$F(b) - F(a).$$

Hence expressions (17) and (15) are always equal to $F(b) - F(a)$, regardless of the choice of subdivision of $[a, b]$. Since expression (15) is a constant, it approaches the limit $F(b) - F(a)$ as the subdivision is chosen finer and finer. But that limit is also $\int_a^b f(x)\,dx$. We summarize our conclusions in the following proposition.

Proposition 1 (Fundamental Theorem of the Calculus): If f is continuous on the closed interval $[a, b]$ and F is a function such that $F'(x) = f(x)$ for every x in $[a, b]$, then $\int_a^b f(x)\,dx$ exists and is equal to $F(b) - F(a)$.

If f is continuous on $[a, b]$ and if $f(x) \geq 0$ for all x in $[a, b]$, then we can set $A(w)$ equal to the area bounded by the graph of f, the x-axis, and the vertical lines with equations $x = a$ and $x = w$ (Figure 6–15), for each

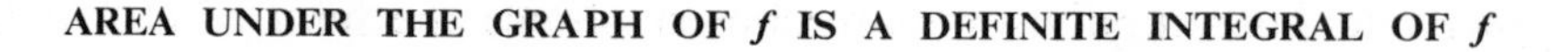

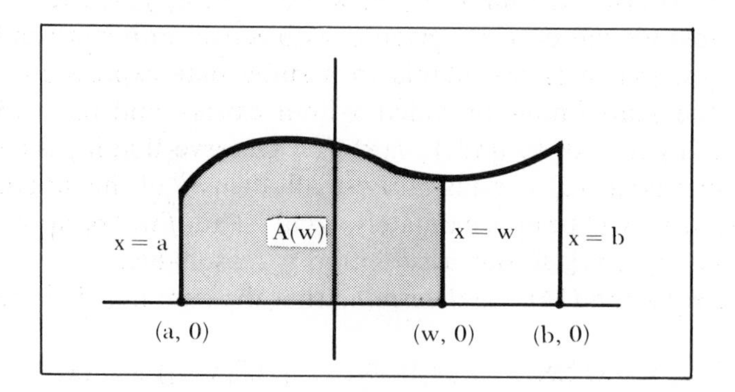

Figure 6–15 Let f be continuous on $[a, b]$ and $f(x) \geq 0$ for all x in $[a, b]$. Set $A(w)$ equal to the area bounded by the graph of f, the x-axis, and the vertical lines $x = a$ and $x = w$. Then $A(w) = \int_a^w f(x)\,dx$.

w in $[a, b]$. $A(w)$ also equals $\int_a^w f(x)\,dx$ according to the geometric interpretation of the definite integral. More generally, we let $A(w) = \int_a^w f(x)\,dx$ for every w in $[a, b]$ whenever f is continuous on $[a, b]$—even when $f(x)$ is non-positive for all x in $[a, b]$.

Now, using an argument similar to that given in Example 3 of Section 6.1, we find that $A'(w) = f(w)$ for each w in $[a, b]$; hence A is an indefinite integral of f. However, according to the geometric interpretation of the definite integral, $A(w) = \int_a^w f(x)\,dx$. Our conclusions are summed up in the following proposition.

Proposition 2 (Second Form of the Fundamental Theorem of the Calculus): If f is continuous on $[a, b]$ and we set

$$A(w) = \int_a^w f(x)\, dx,$$

then $A(w)$ is defined for every point of $[a, b]$; moreover, $A'(w) = f(w)$ for each w in $[a, b]$.

We now have at least partial answers to two very important questions. First, what conditions must f satisfy in order for $\int_a^b f(x)\, dx$ to exist? We have concluded that it suffices for f to be continuous at each point of $[a, b]$. It is in fact true that $\int_a^b f(x)\, dx$ will sometimes exist even when f is *not* continuous at every point of $[a, b]$, but we will not investigate these situations in our discussion. We will limit ourselves to the statement that the definite integral will exist for all continuous functions.

Secondly, how do we evaluate $\int_a^b f(x)\, dx$ once we are certain that it exists? We find an indefinite integral F for f; then $\int_a^b f(x)\, dx = F(b) - F(a)$. Does this indefinite integral for f always exist? It will if f is continuous. However, although we know that if f is continuous on $[a, b]$ then $A(w) = \int_a^w f(x)\, dx$ defines an indefinite integral for f for w in $[a, b]$, it may not always be possible to translate this indefinite integral into an easily computable form (see Example 5). We now present some examples of the application of Propositions 1 and 2.

Example 4: We evaluate

$$\int_{-1}^{2} (x^3 - e^x)\, dx. \tag{18}$$

Since $f(x) = x^3 - e^x$ is continuous on $[-1, 2]$, we are certain that this integral exists. The function $F(x) = (x^4/4) - e^x$ is an indefinite integral for f (this can be checked by differentiating F and finding the result to be f). We compute the value of (18): $F(2) - F(-1) = (2^4/4 - e^2) - ((-1)^4/4 - e^{-1}) = 4 - 1/4 - e^2 + e^{-1} = 15/4 - e^2 + e^{-1}$.

Note that $F(x) = x^4/4 - e^x + c$ is an indefinite integral of $f(x) = x^3 - e^x$ regardless of the value of c. However, c drops out of the computation of $F(b) - F(a)$ anyway (cf. the opening part of Example 1 of this chapter); therefore, we might as well, for convenience, use $x^4/4 - e^x$, the form of F with $c = 0$, when we evaluate $\int_a^b f(x)\, dx$.

We recall that a function is a rule, phrase, or relationship which assigns a unique real number, the function value, to each number in its domain of definition. Generally, the functions encountered thus far have been defined by means of algebraic equations—for example, $f(x) = x^2$—but this does not always have to be the means by which a function is defined. The example below illustrates how a function can be defined using a definite integral.

Example 5: We will now find a function F defined on $[1, 2]$ whose derivative is the function $f(x) = e^x/x$. We have not yet encountered a readily computable function F whose derivative is f. Nevertheless, f is continuous on $[1, 2]$; hence Proposition 2 indicates that if we set

$$F(w) = \int_1^w (e^x/x)\, dx,$$

then $F(w)$ will be defined for each w in $[1, 2]$ and $F'(w)$ will equal $f(w) = e^w/w$ for all w in $[1, 2]$. Thus, we are defining the function F by means of a definite integral. We can interpret $F(w)$ geometrically as the area bounded by the graph of f, the x-axis, and the vertical lines $x = 1$ and $x = w$.

It is evident that because the indefinite integral is vital in evaluating definite integrals, we need techniques for finding indefinite integrals. Finding an indefinite integral essentially reverses the process of differentiation, since $\int f(x)\, dx$, the indefinite integral of f, is a function whose derivative is f.

In certain instances a given function is readily recognizable as the derivative of another function. For example, if $f(x) = 1/x$, then f is easily recognized as the derivative of $F(x) = \log_e x$; hence $F(x) = \log_e x = \int (1/x)\, dx$. Certain properties of the indefinite integral also follow from the basic properties of derivatives. One important property is illustrated in the following example.

Example 6: Suppose that $h(x) = f(x) + g(x)$. If $F(x) = \int f(x)\, dx$ and $G(x) = \int g(x)\, dx$, then by definition $F'(x) = f(x)$ and $G'(x) = g(x)$. Then $F'(x) + G'(x) = f(x) + g(x) = h(x)$. Because the derivative of the sum of two functions is the sum of the derivatives of the functions (cf. Proposition 3 of Chapter 4), we can say that $F'(x) + G'(x) = (F + G)'(x) = h(x)$. Thus we conclude that $F + G$ is the indefinite integral for h, and we can therefore state that the indefinite integral of the sum of two functions equals the sum of their indefinite integrals:

$$\int (f + g)(x)\, dx = \int f(x)\, dx + \int g(x)\, dx. \tag{19}$$

Equation (19) may help us reduce a complicated integral to a sum of simpler integrals. For example,

$$\int (x^2 + x^{1/2})\, dx = \int x^2\, dx + \int x^{1/2}\, dx.$$

The following list contains a number of indefinite integrals. To cover the general case, we include the constant term in the indefinite integral. The constant term should be included when one is solving a differential equation or when the most general form of the indefinite integral is desired; it may be omitted when the indefinite integral is used to evaluate definite integrals. The accuracy of each indefinite integral can be checked directly by differentiating it. The more general properties of the indefinite integral, such as equation (19), follow immediately from the basic properties of the first derivative.

$$\int kf(x)\, dx = k\int f(x)\, dx, \tag{20}$$

where k is a constant.

$$\int (k_1f(x) + k_2g(x))\, dx = k_1\int f(x)\, dx + k_2\int g(x)\, dx, \tag{21}$$

where k_1 and k_2 are constants. Equation (21) is a combination of equations (19) and (20).

$$\int 0\, dx = 0 + c, \tag{22}$$

where c is a constant.

$$\int k\, dx = kx + c, \tag{23}$$

where k is a constant.

$$\int x^n\, dx = x^{n+1}/(n + 1) + c, \quad \text{if} \quad n \neq -1 \tag{24}$$

$$\int x^{-1}\, dx = \log_e x + c \tag{25}$$

$$\int e^x\, dx = e^x + c \tag{26}$$

$$\int a^x\, dx = a^x/\log_e a + c, \tag{27}$$

where a is a positive real number.

The following indefinite integrals can be obtained by methods that we will not explain. Their accuracy can, of course, be checked by differentiating them. Let a be a positive real number.

$$\int \frac{dx}{x^2 - a^2} = \frac{1}{2a} \log_e \left(\frac{x - a}{x + a}\right) + c, \quad \text{for} \quad x^2 > a^2 \tag{28}$$

$$\int \frac{dx}{a^2 - x^2} = \frac{1}{2a} \log_e \left(\frac{a + x}{a - x}\right) + c, \quad \text{for} \quad a^2 > x^2 \tag{29}$$

$$\int (x^2 \pm a^2)^{-1/2}\, dx = \log_e (x + (x^2 \pm a^2)^{1/2}) + c \tag{30}$$

$$\int (x^2 \pm a^2)^{1/2}\, dx = (x/2)(x^2 \pm a^2)^{1/2} \pm (a^2/2) \log_e (x + (x^2 \pm a^2)^{1/2}) + c \tag{31}$$

A more complete table of indefinite integrals can be found in the Appendix.

Example 7: We now evaluate

$$\int (x^2 + 4)^{1/2}\, dx.$$

This indefinite integral has the same general form as equation (31) if we use the plus sign in (31) and let $a = 2$. The desired indefinite integral, the right side of equation (31), then becomes

$$(x/2)(x^2 + 4)^{1/2} + 2 \log_e (x + (x^2 + 4)^{1/2}) + c.$$

Sometimes a given indefinite integral must be modified algebraically before a form can be found in a table of integrals to evaluate it.

Example 8: We now compute

$$\int dx/(3x^2 - 7). \tag{32}$$

As it stands, this indefinite integral does not have the form of any integral in the above list. However, it seems to resemble equation (28). If we divide both numerator and denominator of equation (32) by 3, and then apply equation (20), we obtain

$$\int \frac{dx}{3x^2 - 7} = \int \frac{(1/3)\, dx}{x^2 - (7/3)} = (1/3) \int \frac{dx}{x^2 - (7/3)} \cdot \tag{33}$$

Since the last integral in equation (33) now has the form of equation (28) with $a = (7/3)^{1/2}$, we evaluate integral (32) as

$$\left(\frac{1}{2(7/3)^{1/2}}\right) \log_e \left(\frac{x - (7/3)^{1/2}}{x + (7/3)^{1/2}}\right) + c.$$

A number of indefinite integrals contain trigonometric and inverse trigonometric functions, but such functions will not be discussed in this book. Any of several standard calculus texts may be consulted for an explanation of the basic properties of these functions. We will continue our discussion about finding indefinite integrals in the next section.

Exercises

ROUTINE

1. Solve each of the following integrals.
   a) $\int (x^2 + 1)\, dx$
   b) $\int (x^2 + e^x + 1)\, dx$
   c) $\int_0^1 (x^2 + e^x + 1)\, dx$
   d) $\int_{-1}^1 e^x\, dx$
   e) $\int_{-1}^1 2^x\, dx$
   f) $\int (x^2 - 7)^{-1/2}\, dx$
   g) $\int (x^{0.1} - x^{-0.1})\, dx$
   h) $\int 2(x^2 + 1)(2x)\, dx$
   i) $\int (x^2 + 1)x\, dx$
   j) $\int 3x^{-1}\, dx$
   k) $\int dx/(x + 1)$
   l) $\int (x - 1)\, dx/x$
   m) $\int dx/(2x^2 - 32)$ [Hint: Divide numerator and denominator by 2.]
   n) $\int_{-1}^0 dx/(2x^2 - 32)$
   o) $\int (4x^2 - 9)^{1/2}\, dx$ [Hint: Express $4x^2 - 9$ as $4(x^2 - 9/4)$.]
   p) $\int_3^4 (4x^2 - 9)^{1/2}\, dx$
   q) $\int_3^4 (4x^2 + 9)^{1/2}\, dx$
   r) $\int (x^{0.9} - x^{-1/8} + e^{-1})\, dx$ [Warning: Is e^{-1} a variable or a constant?]

2. What is the distance traveled by a car from $t = 2$ to $t = 3$ if the car travels with the velocity of each of the functions below?
   a) $v(t) = t^{1/2}$
   b) $v(t) = t$
   c) $v(t) = t^2$
   d) $v(t) = t^3$

THEORETICAL

3. Consider the sum

$$f(x_1)(x_1 - x_0) + f(x_2)(x_2 - x_1) + \cdots + f(x_n)(x_n - x_{n-1})$$

from Definition 1 for the definite integral.
   a) What sign does each term of the form $x_i - x_{i-1}$ have? This implies that the sign of $f(x_i)(x_i - x_{i-1})$ is the same as the sign of $f(x_i)$.
   b) If $f(x_i) < 0$, sketch the rectangle whose area is $f(x_i)(x_i - x_{i-1})$. Does this rectangle lie above or below the x-axis?

c) The area we are considering in relation to the definite integral is *signed* area; that is, area above the x-axis will be positive, while area below the x-axis will be negative. Sketch the area represented by the integral $\int_{-1}^{1} x^3\, dx$.

d) Interpret geometrically why the integral in part (c) is equal to 0. What is the value of the area represented by $\int_{-1}^{1} x^3\, dx$ which lies below the x-axis? What is the value of this area which lies above the x-axis?

4. We call a and b in the expression $\int_a^b f(x)\, dx$ the *limits of integration.* Justify the definition represented in

$$\int_b^a f(x)\, dx = -\int_a^b f(x)\, dx.$$

More specifically:

a) Consider equation (10) in Definition 1 of this chapter. If we integrate from b to a, we should reverse the roles of the right and left end points of each subdivision. Rewrite equation (10), interchanging x_i and x_{i-1}.

b) The sum found in part (a) is simply -1 times equation (10). What is the limit of that sum as the subdivisions are chosen finer and finer?

5. Justify the following statement: If $\int_a^b f(x)\, dx$ exists and c is a point of $[a, b]$, then $\int_a^b f(x)\, dx = \int_a^c f(x)\, dx + \int_c^b f(x)\, dx$. For example, $\int_0^1 x^2\, dx + \int_1^2 x^2\, dx = \int_0^2 x^2\, dx$.

6.3 SUBSTITUTION OF VARIABLE. INTEGRATION BY PARTS

More often than not, an integral will not have a form exactly corresponding to one found in a table of integrals. In such a case, however, it is often possible to transform the given integral into an equivalent form which does correspond to an integral in the table. One of the most common methods of changing the form of an integral is the so-called *substitution of variable.* This method is essentially the Chain Rule in reverse.

If $h(x) = F(g(x))$, then we can apply the Chain Rule:

$$h'(x) = F'(g(x))g'(x). \tag{34}$$

Therefore, an expression of the form (34) will have an indefinite integral

$$h(x) = F(g(x)),$$

since we differentiate this function to arrive at (34).

Example 9: Consider $h'(x) = 2(x^3 - 1)(3x^2)$. If we let $g(x) = x^3 - 1$ and $F(x) = x^2$, then $h'(x) = F'(g(x))g'(x)$. Therefore h, the indefinite integral of h', is defined by $h(x) = F(g(x)) = (x^3 - 1)^2 + c$, where c is a constant.

How can we recognize that our given problem is similar to equation (34), and how do we find an appropriate F and g to compute the indefinite integral $F \circ g$?

We employ the following method: To find

$$\int f(g(x))g'(x)\,dx, \tag{35}$$

let $y = g(x)$. Find $\int f(y)\,dy$. Resubstitute $g(x)$ for y after $\int f(y)\,dy$ is found. The resulting expression is the desired indefinite integral. (Relative to the discussion above, $f = F'$ and $F(y) = \int f(y)\,dy$.) We illustrate this technique, called *substitution of variable*, in the following examples.

Example 10: Consider

$$\int xe^{x^2}\,dx. \tag{36}$$

We first note that x is related to the derivative of x^2; if we let $g(x) = x^2$, then $g'(x) = 2x$. If we express integral (36) in the equivalent form

$$(1/2)\int (2x)e^{x^2}\,dx, \tag{37}$$

then integral (36) becomes

$$(1/2)\int e^{g(x)}g'(x)\,dx.$$

To evaluate $\int e^{g(x)}g'(x)\,dx$, we substitute $y = g(x)$ and evaluate $\int e^y\,dy$. Since $\int e^y\,dy = e^y + c$, the method of substitution of variable yields the expression $(1/2)e^{x^2} + c$ for integral (37); note that we have resubstituted $g(x) = x^2$ for y, and we have included the constant term c to arrive at the most general indefinite integral.

Example 11: Next consider

$$\int 3x\,dx/(x^4 - 1). \tag{38}$$

Integral (38) bears some resemblance to equation (28), but x^4 appears instead of x^2 in the denominator and $3x\,dx$ replaces dx in the numerator.

Nevertheless, the similarity between forms (38) and (28) can be made more apparent if we write (38) as

$$\int \frac{3x\,dx}{(x^2)^2 - 1}.$$

If we set $g(x) = x^2$, then this integral assumes the form

$$\int \frac{3(1/2)g'(x)\,dx}{(g(x))^2 - 1} = (3/2)\int \frac{g'(x)\,dx}{(g(x))^2 - 1}. \tag{39}$$

The integral on the right side of equation (39) exhibits the form $\int f(g(x))g'(x)\,dx$, where $f(x) = 1/(x^2 - 1)$; hence it is appropriate to use substitution of variable to find its value. Let $y = g(x)$. In terms of y, the right side of equation (39) is

$$(3/2)\int dy/(y^2 - 1). \tag{40}$$

The integral $\int dy/(y^2 - 1)$ has the same general form as the integral in equation (28); making the appropriate substitutions in that equation, we find

$$\int dy/(y^2 - 1) = (1/2)\log_e\left(\frac{y-1}{y+1}\right) + c.$$

Therefore, integral (40) is equal to

$$(3/2)(1/2)\log_e\left(\frac{y-1}{y+1}\right) + c.$$

Resubstituting $g(x) = x^2$ for y, we conclude that integral (38) is equal to

$$(3/4)\log_e\left(\frac{x^2-1}{x^2+1}\right) + c.$$

Facility in using substitution of variable comes only with practice. Sometimes, finding the appropriate method for evaluating an integral is as much a matter of luck as of skill. For example, the integral

$$\int 3\,dx/(x^4 - 1)$$

bears as much resemblance to equation (28) as does integral (38), yet substitution of variable succeeds with (38) but will fail if applied here.

One should, in fact, attempt to evaluate $\int 3\,dx/(x^4 - 1)$ using substitution of variable to be convinced that the method will not work in this case.

Another useful method for evaluating certain indefinite integrals, or converting them to a form that might be found in a table, is *integration by parts.* Whereas substitution of variable finds its justification in the Chain Rule, integration by parts stems from the formula for the derivative of the product of two functions (cf. Proposition 4 of Chapter 4).

Recall that if $h(x) = f(x)g(x)$, then

$$h'(x) = f(x)g'(x) + f'(x)g(x). \tag{41}$$

Since h is the indefinite integral of h', and since the indefinite integral of the sum on the right side of equation (41) is the sum of the indefinite integrals of its summands, we observe that

$$h(x) = f(x)g(x) = \int f(x)g'(x)\,dx + \int f'(x)g(x)\,dx,$$

or

$$\int f(x)g'(x)\,dx = f(x)g(x) - \int f'(x)g(x)\,dx. \tag{42}$$

Equation (42) constitutes the basis for integration by parts. The following examples will illustrate how this method is applied.

Example 12: Consider the definite integral

$$\int_0^1 xe^x\,dx. \tag{43}$$

To evaluate integral (43), we will first find an indefinite integral $F(x)$ for xe^x using integration by parts; we can then numerically evaluate (43) by computing $F(1) - F(0)$.

Note that xe^x is the product of the two functions x and e^x, and that e^x is also the first derivative of e^x. Therefore, if we set $f(x) = x$ and $g(x) = e^x$, then $\int xe^x\,dx$ assumes the form $\int f(x)g'(x)\,dx$. Applying formula (42), we obtain the result

$$\begin{aligned} F(x) = \int xe^x\,dx &= \int f(x)g'(x)\,dx \\ &= f(x)g(x) - \int f'(x)g(x)\,dx \\ &= xe^x - \int e^x\,dx \\ &= xe^x - e^x + c. \end{aligned}$$

Therefore, integral (43) is found to be $F(1) - F(0) = ((1)e^1 - e^1) - (0e^0 - e^0) = 0 - (-1) = 1$.

Example 13: We will now find the value of the integral

$$\int_4^8 x \log_e x \, dx. \tag{44}$$

We first compute $F(x) = \int x \log_e x \, dx$, and then evaluate integral (44) by finding $F(8) - F(4)$. Since $\int x \log_e x \, dx$ is the integral of the product of two functions, integration by parts is a possible method of evaluation. Now, $\log_e x$ has a simple first derivative but no readily ascertained indefinite integral, while an indefinite integral of x is $x^2/2$. We will therefore set $f(x) = \log_e x$ and $g(x) = x^2/2$. With f and g so defined, $\int x \log_e x \, dx$ takes the form $\int f(x)g'(x) \, dx$. Applying formula (42), we determine that

$$\begin{aligned} F(x) = \int x \log_e x \, dx &= f(x)g(x) - \int f'(x)g(x) \, dx \\ &= (\log_e x)(x^2/2) - \int (1/x)(x^2/2) \, dx \\ &= (1/2)x^2 \log_e x - (1/2)\int x \, dx \\ &= (1/2)x^2 \log_e x - x^2/4 + c. \end{aligned}$$

Therefore, integral (44) is equal to

$$\begin{aligned} F(8) - F(4) &= (1/2)(8^2 \log_e 8 - 8^2/4) - (1/2)(4^2 \log_e 4 - 4^2/4) \\ &= 32 \log_e 8 - 8 \log_e 4 - 6. \end{aligned}$$

We should note that an integral may contain the product of two functions and yet be impossible to compute by integration by parts, or any other means, for that matter. In such cases the integral may be approximated by numerical methods that we will explore briefly in Section 7.2. There are also a considerable number of techniques of integration besides substitution of variable and integration by parts; however, we will not present them in this book. We shall be concerned primarily with converting any integral to one of the standard forms found in a table of integrals. If this is not feasible, then it is doubtful that any method we might present here would be of much use in evaluating the integral.

Exercises

ROUTINE

1. Evaluate each of the following integrals. Each of these integrals can be found by using the short table of Section 6.2, integration by parts, substitution of variable, or a combination of these methods.

a) $\int x(x^2 - 1)\,dx$

b) $\int x^2(x^3 + 1)\,dx$

c) $\int_0^1 x^2(x^3 + 1)^3\,dx$

d) $\int_{-1}^1 e^{e^x}e^x\,dx$

e) $\int \frac{x-1}{x}\,dx$

f) $\int_7^{10} xe^{-x}\,dx$

g) $\int x(x^4 + 9)^{1/2}\,dx$

h) $\int_{10}^{70} x(16x^4 - 81)^{-1/2}\,dx$

i) $\int \frac{\log_e x}{x}\,dx$

j) $\int \frac{\log_e x^{1/2}}{x}\,dx$

k) $\int_3^4 \frac{x-1}{x^2-1}\,dx$

l) $\int \log_e x\,dx$ [Hint: Write $\log_e x$ as $1 \cdot \log_e x$ and use integration by parts.]

m) $\int_0^1 x\,dx/(x + 1)$ [Hint: Write $x/(x + 1)$ as the sum of two fractions.]

n) $\int_0^1 x^2e^x\,dx$ [Hint: Integration by parts may have to be applied more than once.]

o) $\int x^2\,dx/(x + 1)$

p) $\int x\,dx/(x^2 - 4x + 4)$

q) $\int e^x\,dx/(e^x + 1)$

r) $\int \frac{\log_2 x}{x}\,dx$

s) $\int \frac{x^2-1}{x+1}\,dx$

t) $\int \frac{e^{2x}-1}{e^x-1}\,dx$

CHALLENGING

2. The number $n(x)$ of people who score x per cent on a certain test can be estimated by

$$n(x) = -2x + 200,$$

where x is an integer between 10 and 90.

a) Explain why $\int_{30}^{40} n(x)\,dx$ approximates the number of people who score between 30 and 40.

b) Use an appropriate definite integral to estimate the number of people who score more than 60.

6.4 SOME APPLICATIONS OF THE DEFINITE INTEGRAL

Underlying almost every application of the definite integral is the fact that the definite integral constitutes the limit of a special kind of summation process. Suppose that the following conditions are all met:

i) Some quantity Q is sought.

ii) We can find some continuous function f defined on a closed interval $[a, b]$ such that for any subdivision of $[a, b]$, the sum

$$f(x_1)(x_1 - x_0) + f(x_2)(x_2 - x_1) + \ldots + f(x_n)(x_n - x_{n-1})$$

furnishes an approximation to Q; as before, the x_i are the endpoints of the subintervals of the subdivision of $[a, b]$.

iii) The approximation of Q given by this sum becomes more accurate as we choose a finer and finer subdivision.

If conditions (i), (ii), and (iii) are met simultaneously, then

$$Q = \int_a^b f(x)\, dx.$$

Suppose, for example, that $Q(t)$ represents some quantity that depends on time, and that $f(t)$ is the rate at which $Q(t)$ is changing at time t. The total change in $Q(t)$ from $t = a$ to $t = b$ will be $Q(b) - Q(a)$. In many instances, however, $Q(t)$ is not stated explicitly, but we know or can calculate $f(t)$. If f is continuous for t in the closed interval $[a, b]$ and $[a, b]$ is divided into subintervals, then

$$f(t_1)(t_1 - t_0) + f(t_2)(t_2 - t_1) + \ldots + f(t_n)(t_n - t_{n-1})$$

approximates the total change in $Q(t)$ from $t = a$ to $t = b$. This approximation is obtained with the assumption that $Q(t)$ changes at a constant rate over the subinterval $[t_{i-1}, t_i]$. Since this approximation becomes more accurate as the subdivision is chosen finer, conditions (i), (ii), and (iii) are fulfilled in this case. Hence $Q(b) - Q(a)$ will be found by calculating $\int_a^b f(t)\, dt$.

Example 14: In Example 1 we saw that if a vehicle has velocity $v(t)$ at time t—velocity being the rate at which the vehicle's position is changing—then the total distance the vehicle travels from $t = a$ to $t = b$ is

$$\int_a^b v(t)\, dt.$$

Example 15: Water is flowing into a tank at a rate inversely proportional to the number of hours that the water has been flowing plus 1. (Quantity A is *inversely proportional* to quantity B if $A = k/B$, where k is some constant.) If the water is filling the tank initially (at $t = 0$) at 10 cubic feet per hour, how much water flows into the tank in the first hour (from $t = 0$ to $t = 1$)?

According to the information given, if $r(t)$ is the rate of flow at time t, then

$$r(t) = k/(t + 1),$$

where k is some constant, and

$$r(0) = k/1 = 10.$$

Therefore,

$$r(t) = 10/(t + 1),$$

which is the rate at which the amount of water in the tank is changing at time t. This rate is a continuous function for t in [0, 1]. Therefore, the total amount of water in the tank after 1 hour is

$$\int_0^1 10\, dt/(t + 1) = 10 \int_0^1 dt/(t + 1). \qquad (45)$$

The indefinite integral of $f(t) = 1/(t + 1)$ is $\log_e (t + 1)$; therefore, equation (45) becomes

$$10(\log_e 2 - \log_e 1) = 10(\log_e 2 - 0) = 10 \log_e 2 \text{ cubic feet.}$$

The definite integral can also be used when the variable associated with the rate of change of some quantity is not time. Suppose that:

a) $Q(x)$ is a quantity dependent on x.
b) $f(x)$ is the first derivative of $Q(x)$.
c) f is continuous on the closed interval $[a, b]$.

If conditions (a), (b), and (c) are all fulfilled, then

$$Q(b) - Q(a) = \int_a^b f(x)\, dx.$$

This is just a restatement of the Fundamental Theorem of the Calculus. It sometimes happens that it is easier to compute $f(x) = Q'(x)$ than it is to find $Q(x)$ itself. In such instances we can use f and the definite integral to calculate Q.

Example 16: In Example 3 we saw that the area function A—$A(w)$ was the area bounded by the graph of a continuous non-negative function f, the x-axis and the vertical lines $x = 0$ and $x = w$—has f as its first derivative. The function f is specified and we use the definite integral to calculate $A(w)$. In particular,

$$A(w) = \int_0^w f(x)\,dx.$$

Example 17: The *probability distribution function* F of a random variable X is defined by

$F(x) =$ probability that X will assume a value less than or equal to x.

It can be shown that if F is continuous, then the probability that X will have a value somewhere in the interval $[a, b]$ is $F(b) - F(a)$. With regard to many random variables, it is easier from both a theoretical and a computational standpoint to find and deal with the first derivative of the probability distribution function than with the distribution function itself. This first derivative, $F' = f$, is called the *probability density function*. In terms of the density function f, the probability $F(b) - F(a)$ that X will assume a value in $[a, b]$ is $\int_a^b f(x)\,dx$. We observe that since $f = F'$, the above statement is simply the Fundamental Theorem of the Calculus.

Example 18: The outcome of a certain experiment indicates that the density function f for the random variable T, whose value is the length in minutes of a phone call, is given by

$$f(t) = 2e^{-2t},$$

where $t \geq 0$. Thus the probability that any given phone call will last between 1 and 2 minutes is

$$\int_1^2 f(t)\,dt = \int_1^2 2e^{-2t}\,dt = e^{-2} - e^{-4}.$$

It is customary to represent a random variable by a capital letter and specific values of the random variable by a small letter of the same type; thus, the random variable T here assumes a value t. Note that T is actually a function that measures the length of a phone call; $T(p) = t$, where p is a particular phone call and t is the length of p.

In some instances we can approximate discontinuous functions by functions that are continuous (or even differentiable). Then we employ the techniques of the calculus to gather information about the original function. We have already encountered this method in Example 9 of Chapter 5, where we used a continuous function to approximate a function that was defined only for the non-negative integers.

A function is called *discrete* if its domain consists of isolated real numbers (if each point at which the function is defined has no other points close to it for which the function is also defined). Thus, for example, a function defined only for the non-negative integers is discrete. The graph of a discrete function appears in Figure 6–16.

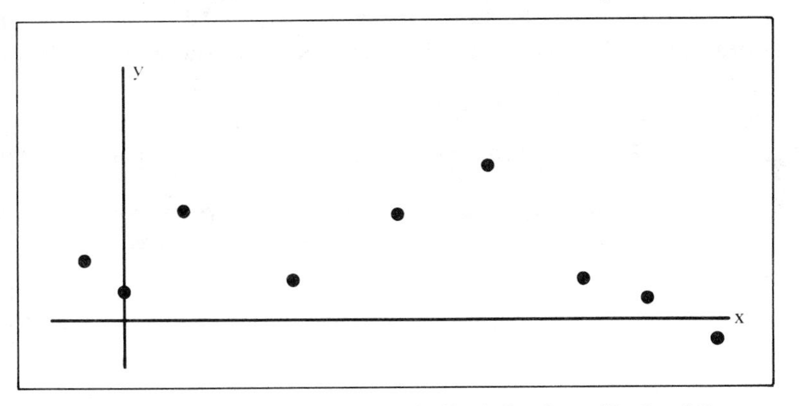

Figure 6–16 The points at which f is defined are "isolated."

Using continuous functions to approximate discrete functions is a particularly powerful tool in probability and statistics. Figure 6–17 illustrates how a continuous function might approximate a discrete function, and in the next example we use such an approximation for computation.

APPROXIMATING A DISCRETE FUNCTION WITH A CONTINUOUS FUNCTION

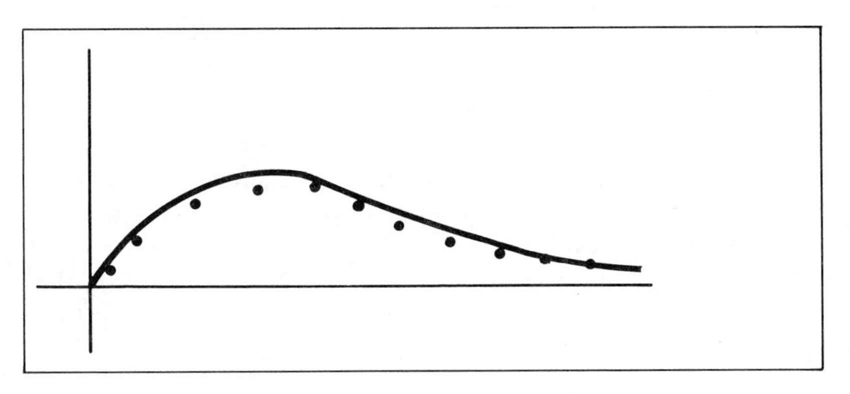

Figure 6–17 The graph of the continuous function (solid line) approximates the graph of the discrete function (dots).

Example 19: A manufacturer prefers to sell upholstered armchairs in quantity rather than one at a time; therefore, he offers discounts when a purchaser buys more than one chair. The first chair costs the buyer $100. The price of the nth chair, where $n \geq 2$, is computed according to the formula

$$P(n) = \text{price of } n^{\text{th}} \text{ chair} = 100 - (1.01)^n,$$

up to a maximum of 50 chairs.

Therefore, $P(n)$ is defined for the integers 2 through 50. We can approximate P by the continuous function $f(x) = 100 - (1.01)^x$, and we can approximate the total cost of 50 chairs using the definite integral as follows.

The rectangle of height $P(2)$, whose base is the interval $[2, 3]$, has area $P(2) \cdot 1 = P(2)$ (Figure 6–18); hence this area is a geometric rep-

AREA UNDER A DISCRETE FUNCTION

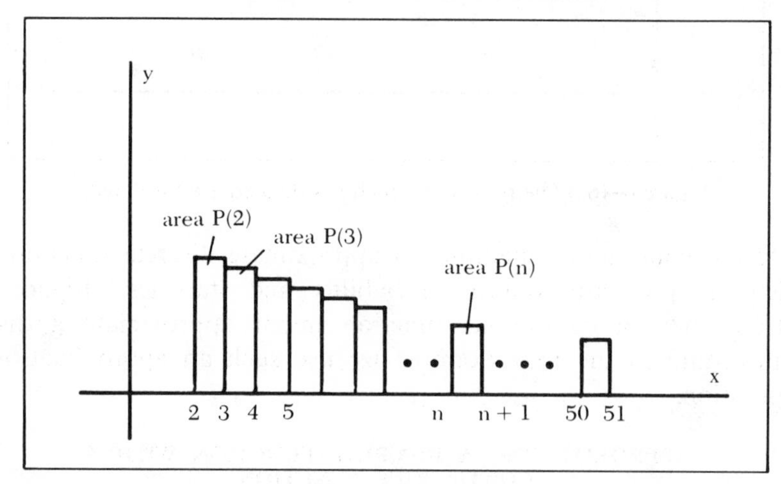

Figure 6–18 The total cost of 50 chairs can be represented as the sum of 50 rectangles in the coordinate plane. The nth rectangle, which represents the cost of the nth chair, has its base on the interval $[n, n + 1]$ and height $P(n)$, the price of the nth chair. We let n be greater than or equal to 2.

resentation of the price of the second chair. In like manner, the rectangle with height $P(n)$, whose base is the interval $[n, n + 1]$, has area $P(n)$. The sum of these rectangles for $n = 2, 3, 4, \ldots, 50$ is the total cost of the second through fiftieth chairs. This same area is approximately the area between the graph of f and the x-axis, between $x = 2$ and $x = 51$ (Figure 6–19). This area is in turn given by the integral

$$\int_2^{51} (100 - (1.01)^x)\, dx. \tag{46}$$

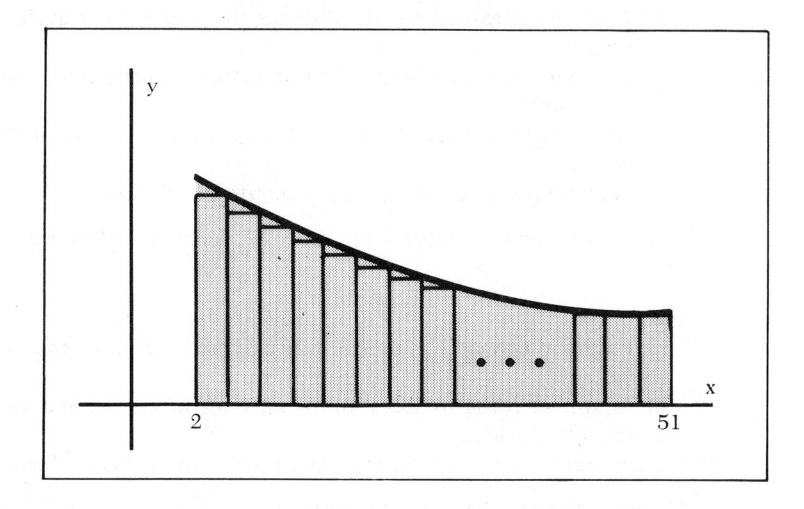

Figure 6–19 The sum of the areas of the rectangles in Figure 6–18 is approximated by the area bounded by the graph of $y = 100 - (1.01)^x$, the x-axis, and the vertical lines $x = 2$ and $x = 51$. The area is the integral in equation (46).

Therefore, integral (46) can be used to approximate $P(2) + P(3) + \ldots + P(50)$. If we add to this the cost of the first chair, then the total cost of 50 chairs is approximately

$$100 + \int_2^{51} (100 - (1.01)^x)\, dx.$$

The applications of the integral given in this section only illustrate the enormous potential usefulness of the integral in practical situations. A catalog of all possible applications of the integral would require a large book of its own. Our purpose has been to present the essential concepts of the integral so that you can apply the integral in your own field of interest and understand its application in other fields.

Exercises

ROUTINE

1. A rocket is accelerating at a rate of 10 feet per second per second.
   a) Calculate the rocket's increase in velocity in the first 10 seconds of its flight (from $t = 0$ to $t = 10$).

b) If the rocket's initial velocity is 0, express the velocity as a function of time.
c) Find an expression for the distance the rocket has traveled after time t.
d) Compute the total distance the rocket travels in the first 20 seconds of its flight.
e) How long will it take for the rocket to attain a velocity of 100 feet per second?
f) How long will it take for the rocket to go 500 feet?

2. A certain random variable X has a density function f given by

$$f(x) = e^{-x},$$

where $x \geq 0$.
a) Find the probability that X will assume a value in the interval $[0, 2]$.
b) Find a such that X has a probability 0.5 of assuming a value in the interval $[0, a]$.
[Cf. Examples 17 and 18, which indicate how to solve this problem.]

3. Compute the area between the graph of

$$y = e^x/(e^x + 1)$$

and the x-axis, between $x = 1$ and $x = 3$. Sketch this area.

4. A company computes its profits on cassette tapes at the rate of

$$10 + 0.001(n - 1)$$

for the nth tape purchased by a single buyer. Approximate this profit function by a continuous function and use the continuous function to estimate the total profit realized if one purchaser buys 100 tapes.

5. Gas is being pumped into a balloon at a rate of

$$r(t) = 10 - 0.05t^2 \text{ cubic feet per second}$$

until the rate reaches 0. Find the total volume of gas in the balloon when the pumping stops.

6. Using an appropriate definite integral, approximate

$$1 + 1/2^2 + 1/3^2 + 1/4^2 + \ldots + 1/100^2.$$

7. A boulder is dropped from the edge of a cliff 5000 feet above the canyon floor. Gravity causes the falling boulder to accelerate at the rate of 32 feet per second per second.
a) How fast is the boulder falling after 3 seconds?
b) How far has it fallen after 4 seconds?
c) How long does it take for the boulder to reach the canyon floor?

8. A company estimates that it costs $(100 - n^{0.1})$ dollars to produce the nth rocking chair it makes. The company also estimates that if it sells a rocking chair for $100 + x$ dollars, where x is some positive number, then $600 - (4 + x)^{0.05}$ chairs will be sold. If the company produces 500 chairs to sell for \$150 each, use the definite integral to estimate the total profit of the company after all the chairs are sold.

CHALLENGING

9. Estimate each of the following sums using the definite integral.
   a) $1 + 1/2 + 1/3 + \ldots + 1/999 + 1/1000$
   b) $1 + 2 + 3 + \ldots + 999 + 1000$
   c) $1^2 + 2^2 + 3^2 + \ldots + 99^2 + 100^2$
   d) $e^{-1} + e^{-2} + e^{-3} + \ldots + e^{-9} + e^{-10}$

10. A study on the amount of time it takes a child to trace his way through a maze concludes that the time t_n the child will take on his n^{th} try is equal to $t_1 n^{-1/2}$; t_1 is the time the child takes on his first try. If a child takes 10 minutes on his first try, how many minutes will he take on the 9th trial? Use the definite integral to estimate the total amount of time the child takes to do 50 trials.

Review of Chapter 6

If f is a function defined on the closed interval $[a, b]$, then the *definite integral* $\int_a^b f(x)\,dx$ of f on $[a, b]$ is defined to be the limit of the sums of the form

$$f(x_1)(x_1 - x_0) + f(x_2)(x_2 - x_1) + \ldots + f(x_n)(x_n - x_{n-1}),$$

where $a = x_0 \leq x_1 \leq x_2 \leq \ldots \leq x_{n-1} \leq x_n = b$; the sums approach a limit as we subdivide $[a, b]$ into more and smaller subintervals. If f is continuous on $[a, b]$, then the limit of the sums, and hence $\int_a^b f(x)\,dx$ will always exist. We interpret $\int_a^b f(x)\,dx$ geometrically as the area in the coordinate plane bounded by the x-axis, the graph of f, and the lines with equations $x = a$ and $x = b$.

If f is continuous on $[a, b]$ and if F is a function such that $F'(x) = f(x)$ for all x in (a, b), then $\int_a^b f(x)\,dx = F(b) - F(a)$. This fact makes it desirable to derive methods by which, given a function f, we can obtain a function F such that $F' = f$; we call such a function F an *indefinite integral* of f.

Many indefinite integrals (and thus many definite integrals) can be found using the methods of substitution of variable, integration by parts, matching the integral with an integral obtained from a table, or a combination of these methods. A short table of integrals appears in the Appendix.

A quantity can often be calculated by means of a definite integral if the quantity can be approximated by sums such as that shown above. For example, if $r(t)$ is the rate at which a quantity $Q(t)$ is changing, then $\int_a^b r(t)\,dt$ will be the total change of $Q(t)$ from $t = a$ to $t = b$.

REVIEW EXERCISES

1. Evaluate each of the following, using any of the techniques discussed in this chapter.

a) $\int_{-}^{1} x^2\,dx$

b) $\int_{-1} x^3\,dx$

c) $\int xe^{x^2}\,dx$

d) $\int (x + 1) \log_e x\,dx$

e) $\int_3^4 xe^{-x}\,dx$

f) $\int (x + 1)\,dx/(x^2 + 2x + 1)$

g) $\int_4^9 e^{\log_e x}\,dx$

h) $\int (1/x)(\log_e (\log_e x))\,dx$

2. Find the total distance traveled by an object from time $t = 0$ to $t = 2$ if the object is traveling with velocity $v(t) = 3t^2 - 1$ at time t.

3. A firm estimates that t years after its founding, its annual rate of growth in millions of dollars of sales can be estimated by e^{-t}. If the firm began with sales of \$4 million, what are the sales of the company one year after its founding?

Vocabulary

improper integral
Trapezoidal Rule
Simpson's Rule

Chapter 7

Further Methods and Applications of Integration

7.1 IMPROPER INTEGRALS

Example 1: Consider the function

$$f(x) = 1/x^2, \qquad x \geq 1.$$

The graph of f appears in Figure 7–1. Note first that as x increases without bound, $f(x)$ approaches the value 0. In fact, if $f(x)$ is to approximate 0 with accuracy $p > 0$, we need merely select x greater than $1/\sqrt{p}$, since $x > 1/\sqrt{p}$ implies

$$f(x) = 1/x^2 < 1/(1/\sqrt{p})^2 = 1/(1/p) = p$$

(see Figure 7–2).

Since $f(x)$ approximates 0 as accurately as we wish provided x is large enough, we can reasonably say that $f(x)$ has 0 as a "limit" as x increases without bound. Instead of letting x approach some specific real number a, we allow x to grow arbitrarily large; as x increases, $f(x)$ approaches 0. Although $f(x)$ appears to have a limit, our previous definition of limit does not apply in this case. Hence, this new type of limit requires a new and more precise definition, which we will formulate later in this section.

Next, consider the area bounded by the graph of f, the x-axis, and the line $x = 1$ (Figure 7–3). Using only the given information, we cannot determine whether the area in question is finite or infinite. Indeed, at present we have no way of finding this area, since it is not that of a regular geometric figure for which we can apply a standard formula. Nor can this area be computed using some appropriate definite integral, because the area is not bounded on the right by a vertical line. However, if $1 < w$,

GRAPH OF $f(x) = 1/x^2, x \geq 1$

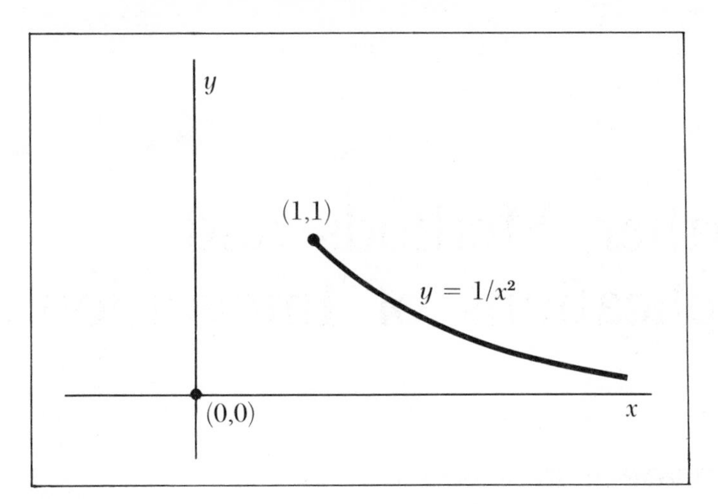

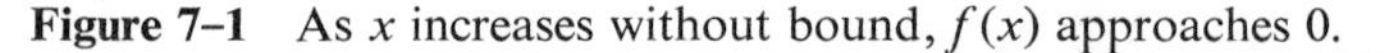

Figure 7–1 As x increases without bound, $f(x)$ approaches 0.

CRITERION FOR ACCURACY OF APPROXIMATION

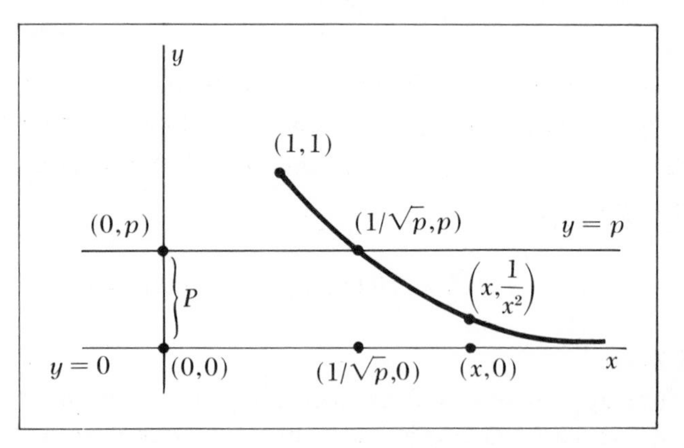

Figure 7–2 For $f(x)$ to approximate 0 with accuracy $p > 0$, x must be greater than $1/\sqrt{p}$. If $x > 1/\sqrt{p}$, then the point $(x, 1/x^2)$ on the graph of f lies between the x-axis and the line $y = p$; hence the distance from $(x, 1/x^2)$ to $(x, 0)$ is less than p.

then the area bounded by the graph of f, the x-axis, and the lines $x = 1$ and $x = w$ is given by

$$\int_1^w (1/x^2)\, dx = 1 - (1/w) \tag{1}$$

(Figure 7–4). In other words, if the area is bounded on the right by the vertical line $x = w$, then the definite integral (1) is used to compute the area. Observe, though, that as w increases, the area bounded by the graph

AN AREA THAT CANNOT BE APPROXIMATED BY A DEFINITE INTEGRAL

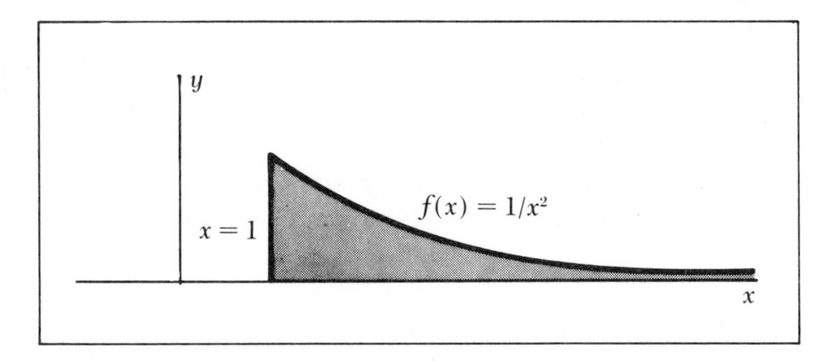

Figure 7–3 The area bounded by the graph of $f(x) = 1/x^2$, the x-axis, and the line $x = 1$ cannot be computed using a definite integral, since it is not bounded on the right by a vertical line. It is not clear whether the area is finite or infinite.

A WAY OF APPROACHING THE PROBLEM IN FIGURE 7–3

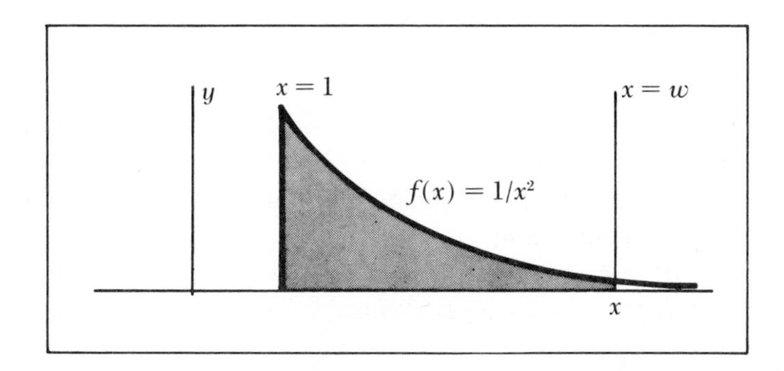

Figure 7–4 The area bounded by the graph of f, the x-axis, and the vertical lines $x = 1$ and $x = w$ forms part of the area in Figure 7–3. As the value of w increases, this area approximates more and more closely the unbounded area. As w increases without bound, the area approaches 1 as a limit.

of f, the x-axis, and the lines $x = 1$ and $x = w$ includes more and more of the area in Figure 7–3. Moreover, as w increases without bound, the area in Figure 7–4, as calculated by integral (1), tends to 1 as a "limit" (since $1/w$ approaches 0 as a "limit" as w becomes arbitrarily large). Once again we are dealing with a limit of a type not formally discussed in previous chapters; moreover, we are examining an area (in Figure 7–3) to which a numerical value can reasonably be assigned, even though there is no definite integral that properly expresses this area.

We will now extend the definitions of the definite integral and the limit of a function to enable us to handle the type of situation presented in Example 1. We first extend the notion of the limit.

Definition 1: *A function f is said to have* **a limit L as x goes to infinity** *if:*

i) *$f(x)$ is defined for values of x chosen as large as we please (meaning that for any number M there exists $x > M$ such that $f(x)$ is defined), and*

ii) *for any $p > 0$, there is a number K such that for any $x > K$ for which $f(x)$ is defined, $|f(x) - L| < p$.*

We can also express condition (ii) this way: It is possible to make $f(x)$ approximate L as closely as we wish by choosing x large enough. Note, too, that the number K in condition (ii) depends on p; generally, the smaller p is, the larger K must be in order that $f(x)$ will approximate L with accuracy p. We can also define the notion of limit for the case in which x is *decreasing* without bound.

Definition 2: *A function f is said to have* **a limit L as x goes to negative infinity** *if:*

i) *for any given number M, there exists $x < M$ such that $f(x)$ is defined, and*

ii) *for any $p > 0$, there is a number K such that for any $x < K$ for which $f(x)$ is defined, $|f(x) - L| < p$.*

NOTATION: We denote *the limit of f as x goes to infinity* by

$$\lim_{x \to \infty} f(x);$$

the limit of f as x goes to negative infinity is denoted by

$$\lim_{x \to -\infty} f(x).$$

From Example 1 we obtain $\lim_{x \to \infty} (1/x^2) = 0$ and $\lim_{x \to \infty} (1/w) = 0$. The following examples further illustrate the limits defined above. We first note, however, that the types of limits defined in Definitions 1 and 2 obey essentially the same rules as do the limits considered earlier in our discussion. For example,

$$\lim_{x \to \infty} f(x) + \lim_{x \to \infty} g(x) = \lim_{x \to \infty} (f + g)(x), \tag{2}$$

provided both limits on the left side of equation (2) exist and $f(x)$ and $g(x)$ are both defined for the same values of x if x is large enough.

Example 2: Consider the function

$$g(x) = x/(x + 1).$$

Since g is defined for all real numbers except -1, $g(x)$ is defined for arbitrarily large and arbitrarily small values of x. Dividing both numerator and denominator of $x/(x + 1)$ by x, we obtain

$$g(x) = \frac{1}{1 + (1/x)}. \tag{3}$$

Now $\lim_{x\to\infty} (1/x) = 0$, and $\lim_{x\to-\infty} (1/x)$ is also 0. Therefore, it follows that

$$\lim_{x\to\infty} g(x) = \lim_{x\to-\infty} g(x) = 1/(1 + 0) = 1.$$

Note how these limits appear in the graph of g (Figure 7–5).

LIMITS AS $x \to \pm\infty$

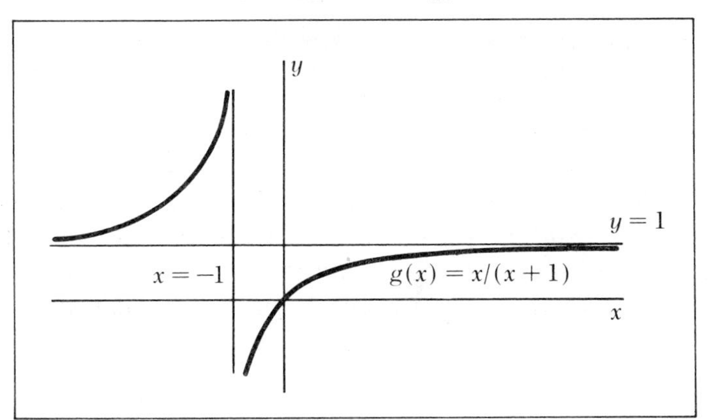

Figure 7–5 If $g(x) = x/(x + 1)$, then $\lim_{x\to\infty} g(x) = \lim_{x\to-\infty} g(x) = 1$. In the graph of g these limits appear as the graph of g approaching the line $y = 1$ as x increases without bound and as x decreases without bound.

Example 3: The function

$$f(x) = x^2$$

has no limit as x goes to either infinity or negative infinity, for in both cases the values of $f(x)$ become arbitrarily large and do not approach any specific real number.

$\lim_{x \to \infty} g(x)$ EXISTS BUT $\lim_{x \to -\infty} g(x)$ DOES NOT

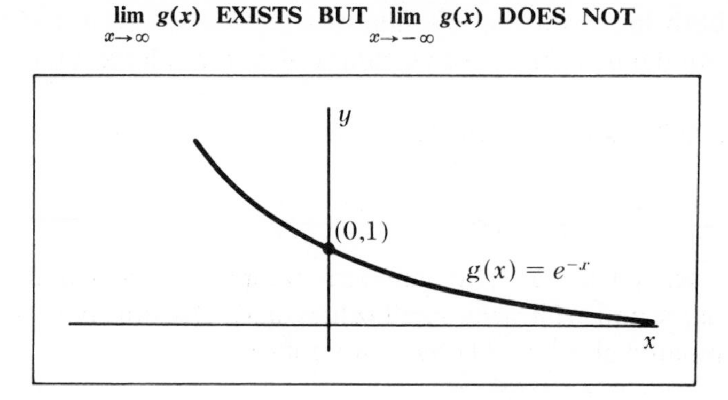

Figure 7–6 If $g(x) = e^{-x}$, then $\lim_{x \to \infty} g(x) = 0$, while $\lim_{x \to -\infty} g(x)$ fails to exist. The graph of g approaches the x-axis as x increases without bound; it does not approach any line as a limit as x decreases without bound.

Example 4: Consider the function

$$g(x) = e^{-x}.$$

From the graph of g (Figure 7–6) it is evident that $\lim_{x \to \infty} g(x) = 0$, while g has no limit as x approaches negative infinity.

Having extended the notion of the limit, we can now extend the definition of the definite integral.

Definition 3: *Suppose that a function f is defined at each point of the open interval (a, b). We will allow a to represent $-\infty$ and b to represent ∞. Recall that $(a, \infty) = \{x \mid a < x\}$, while $(-\infty, \infty)$ is simply the entire set of real numbers. Suppose further that if $a < u < v < b$, then*

$$\int_u^v f(x)\, dx$$

*exists (f is integrable on every closed subinterval of (a, b)). We define the **improper integral of f over (a, b)** to be*

$$\lim_{\substack{u \to a \\ v \to b}} \int_u^v f(x)\, dx. \tag{4}$$

*If these limits exist, we say that the improper integral **converges;** otherwise we say that the integral **diverges.***

In the event that f is integrable over the closed interval $[a, b]$, the improper integral of f over (a, b) can be proved to be the same as $\int_a^b f(x)\,dx$. We therefore let $\int_a^b f(x)\,dx$ denote both the definite integral of f over $[a, b]$, and the improper integral of f over (a, b). The context in which the integral is used determines whether it is to be considered *proper*, and thus in conformity with the ideas and methods of the previous chapter, or *improper*, and hence to be evaluated according to expression (4).

In the following examples we examine several important improper integrals. We first recognize, however, that improper integrals obey essentially the same rules as do proper integrals; for example,

$$\int_a^b f(x)\,dx + \int_a^b g(x)\,dx = \int_a^b (f + g)(x)\,dx, \tag{5}$$

provided both integrals on the left side of equation (5) converge.

Example 5: The integral that expresses the area in Figure 7–4 (Example 1) is

$$\int_1^\infty (1/x^2)\,dx.$$

By definition, this integral is expressed as

$$\lim_{u\to\infty} \int_1^u (1/x^2)\,dx = \lim_{u\to\infty} (1 - (1/u)) = 1.$$

The definition of a proper integral is concerned with functions defined on a closed interval. We now extend the meaning of *interval* to include not only closed intervals, but open intervals, half-open intervals, the entire set of real numbers, and sets of the form (a, ∞), $[a, \infty)$, $(-\infty, a)$, and $(-\infty, a]$. The notion of improper integral enables us to talk about the integral of a function not merely over a closed interval, but over any interval.

Example 6: The concept of a *density function* is central to probability and statistics. Two features which characterize a *probability density function* f with an interval I as its *admissible range* are (i) $f(x) \geq 0$ for each x in I, and (ii) the integral, either proper or improper, of f over I exists and is equal to 1. Since $f(x) = 1/x^2$ has properties (i) and (ii) relative to the interval $(1, \infty)$, f is a possible probability density function with $(1, \infty)$ as its admissible range.

Now consider again the function

$$g(x) = e^{-x}.$$

We will integrate g over the interval $(0, \infty)$; that is, we will calculate

$$\int_0^\infty e^{-x}\,dx. \tag{6}$$

Expression (6) is represented geometrically by the area bounded by the graph of g, the x-axis, and the y-axis (but not including the y-axis, as shown in Figure 7–7). For $0 < u < v < \infty$ we compute

$$\int_u^v e^{-x}\,dx = -e^{-v} + e^{-u} = e^{-u} - e^{-v}. \tag{7}$$

Now $\lim_{u\to 0} e^{-u} = 1$, while $\lim_{v\to\infty} e^{-v} = 0$. Therefore, the limit of equation (7) as u goes to zero and v goes to infinity is $1 - 0 = 1$. Since $e^{-x} \geq 0$ for each x in $(0, \infty)$, we see that g is a probability density function with admissible range $(0, \infty)$.

GEOMETRIC INTERPRETATION OF THE IMPROPER INTEGRAL $\int_0^\infty e^{-x}\,dx$

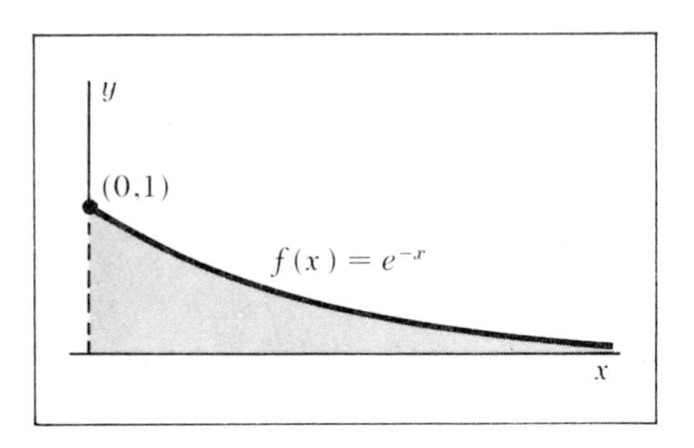

Figure 7–7 The area bounded by the graph of $g(x) = e^{-x}$ and the x-axis for x in the interval $(0, \infty)$ is equal to the improper integral $\int_0^\infty e^{-x}\,dx$.

Observe that integral (6) must be evaluated using two limits rather than simply one, as in Example 5, because we are dealing here with the interval $(0, \infty)$, which does not contain either endpoint. The integral in equation (7) can be computed if $0 < u < v < \infty$, but the definite integral $\int_0^v e^{-x}\,dx$ is meaningless because 0 is not a point of the interval over which the integral is calculated.

Example 7: Consider once again the function

$$f(x) = 1/x^2.$$

We will investigate the improper integral

$$\int_0^1 (1/x^2)\,dx. \tag{8}$$

This integral is improper because f is not defined at 0: one of the limits of integration is a point at which the function value does not exist. For $0 < u < 1$, we obtain

$$\int_u^1 (1/x^2)\, dx = 1 - (1/u). \tag{9}$$

Now, the limit of equation (9) as u approaches 0 does not exist; consequently, the improper integral (8) does not converge. This can be interpreted geometrically to mean that the area bounded by the graph of f, the x-axis, and the y-axis is not finite (Figure 7–8).

A DIVERGENT IMPROPER INTEGRAL

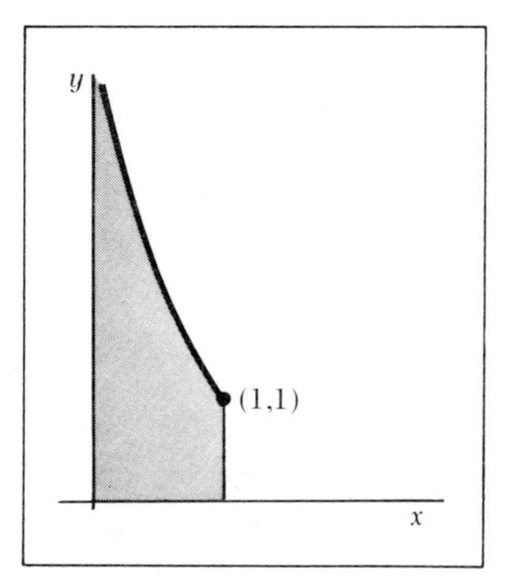

Figure 7–8 The area associated with the improper integral $\int_0^1 (1/x^2)\, dx$ is not finite since the improper integral diverges.

Exercises

ROUTINE

1. Determine which of the following limits exist, and compute the limit if it exists. Review the rules concerning limits (see Section 2.2). These rules are equally valid here.

a) $\lim_{x\to\infty} (1/x)^2$

b) $\lim_{x\to\infty} (7/x)^2$

c) $\lim_{x\to\infty} (-3/x)^2$

d) $\lim_{x\to\infty} e^{-x/2}$

e) $\lim_{x\to\infty} e^{-x/100}$

f) $\lim_{x\to\infty} 78e^{-78x}$

g) $\lim_{x\to-\infty} e^{\sqrt{x}}$

h) $\lim_{x\to\infty} x^2/(x^2 + 1)$

i) $\lim_{x\to-\infty} x^2/(x^2 + 1)$

j) $\lim_{x\to-\infty} \sqrt{x}/(x + 1)$

k) $\lim_{x\to\infty} \sqrt{x}/(x + 1)$

l) $\lim_{x\to\infty} x/\log_e x$ [Hint: Suppose that this limit exists and is equal to L. What does this assumption imply about the ratio of a number to its natural logarithm?]

2. Sketch the graphs of each of the functions appearing in Exercise 1.

3. Evaluate each of the following improper integrals. If an integral diverges, write *diverges*.

a) $\int_1^\infty x^{-3}\,dx$

b) $\int_1^\infty 3x^{-3}\,dx$

c) $\int_1^\infty -2x^{-3}\,dx$

d) $\int_{-\infty}^0 e^x\,dx$

e) $\int_0^1 x^{-1/3}\,dx$

f) $\int_0^1 x^{-1}\,dx$

g) $\int_0^1 x^{-3}\,dx$

h) $\int_{-\infty}^{-1} x^{-2}\,dx$

i) $\int_0^1 \log_e x\,dx$

j) $\int_0^1 x\,dx/(x^2-1)$

k) $\int_8^\infty x\,dx/(x^2-1)$

l) $\int_8^\infty x^2\,dx/(x^3-1)$

CHALLENGING

4. Indicate the area represented by each convergent improper integral in Exercise 3. Are any of these functions probability functions (see Example 6) with admissible ranges equal to the intervals over which they are being integrated? If not, explain why they are not.

THEORETICAL

5. Justify equation (5) on page 205 for improper integrals.

6. Suppose that the improper integral $\int_a^\infty f(x)\,dx$ converges and that $a < b < \infty$. Show that $\int_a^\infty f(x)\,dx = \int_a^b f(x)\,dx + \int_b^\infty f(x)\,dx$.

7.2 NUMERICAL METHODS OF INTEGRATION

All too often a definite integral encountered in the solution of a practical problem cannot be evaluated using integration by parts, substitution of variable, or even a fairly complete table of integrals. Such integrals are presented in the following examples.

Example 8: The integral

$$\int_1^2 \frac{e^x\,dx}{x} \tag{10}$$

exists because e^x/x is continuous on the closed interval $[1, 2]$, but we cannot evaluate this integral exactly by any methods (including using a table of integrals) introduced thus far.

Example 9: The integral

$$\int_{-1}^{1} (1/\sqrt{2\pi})e^{-x^2/2}\,dx \tag{11}$$

represents the probability that a normally distributed random variable will assume a value in the interval $[-1, 1]$. Since the normal probability distribution occurs frequently in statistics, integrals of this type are of great importance. Yet an exact evaluation of this integral is again impossible by any of the methods discussed so far (or, in fact, by any other means).

In the case of many integrals, including (10) and (11), we can at best only approximate their value. In this section we discuss numerical methods of approximating the value of an integral. Before using any approximation technique, however, one should have reason to believe that the integral being sought does in fact exist. We know, for example, that $\int_a^b f(x)\,dx$ will always exist if f is continuous on $[a, b]$. Virtually all integrals that one is likely to encounter will exist, and all integrals considered in this section will exist. Trying to approximate an integral which does not exist is worse than an exercise in futility, because an "approximation" to a non-existent integral may mislead one into believing that the integral does in fact exist. Not only is the approximation wrong, but it conveys a wrong implication as well.

Recall that $\int_a^b f(x)\,dx$ is actually the limit of certain kinds of sums. Moreover, each term of any such sum can be interpreted as the area of a rectangle in the coordinate plane, in which rectangles lying below the x-axis have negative area and rectangles above the x-axis have positive area. (Review Definition 1 of Chapter 6 and the discussion preceding it.) Consequently, any sum such as expression (10) in Chapter 6 is actually an approximation of $\int_a^a f(x)\,dx$. In particular, such a sum is

$$f(x_0)(x_1 - x_0) + f(x_1)(x_2 - x_1) + \ldots + f(x_{n-1})(x_n - x_{n-1}), \tag{12}$$

where the points $x_0, x_1, \ldots, x_n$ form a subdivision of $[a, b]$.

Recall that $x_0 = a$ and $x_n = b$. Also, if we are certain that $\int_a^b f(x)\,dx$ exists and we want to approximate $\int_a^b f(x)\,dx$ using a sum such as (12), we can choose a subdivision that makes the computation as simple as possible. For instance, by choosing points that are equally spaced in $[a, b]$ (dividing $[a, b]$ into n equal subintervals), we can state that

$$x_1 - x_0 = x_2 - x_1 = \ldots = x_n - x_{n-1} = (b - a)/n,$$

and

$$x_0 = a, \quad x_1 = a + (b - a)/n, \quad x_2 = a + 2(b - a)/n, \ldots,$$
$$x_n = a + n(b - a)/n = b,$$

and, in general,

$$x_i = a + i(b - a)/n, \qquad i = 0, 1, 2, \ldots, n.$$

In this case sum (12) may be written

$$((b - a)/n[f(a) + f(a + (b - a)/n) + \ldots + f(a + (n - 1)(b - a)/n)]. \tag{13}$$

The larger n is, the more exactly sum (13) should approximate $\int_b^a f(x)\,dx$.

Example 10: Consider again integral (10) of Example 8. We will divide the interval $[1, 2]$ into four equal subintervals ($n = 4$), and we will evaluate sum (13) for this integral. The length of each subinterval is $(2 - 1)/4 = 1/4$, and the end points of the subintervals are

$$1, 5/4, 3/2, 7/4, 2.$$

Thus sum (13) becomes

$$\frac{1}{4}\left(e + \frac{e^{5/4}}{5/4} + \frac{e^{3/2}}{3/2} + \frac{e^{7/4}}{7/4}\right). \tag{14}$$

We can then evaluate sum (14)—at least approximately—using logarithms, a table of values of e^x, or a computer.

Expression (13) represents the shaded area in Figure 7–9, while the integral $\int_a^b f(x)\,dx$ corresponds to the area bounded by the graph of f, the x-axis, and the lines $x = a$ and $x = b$. Remember that area above the x-axis has positive sign, while area below the x-axis has negative sign.

GEOMETRIC INTERPRETATION OF EXPRESSION (13)

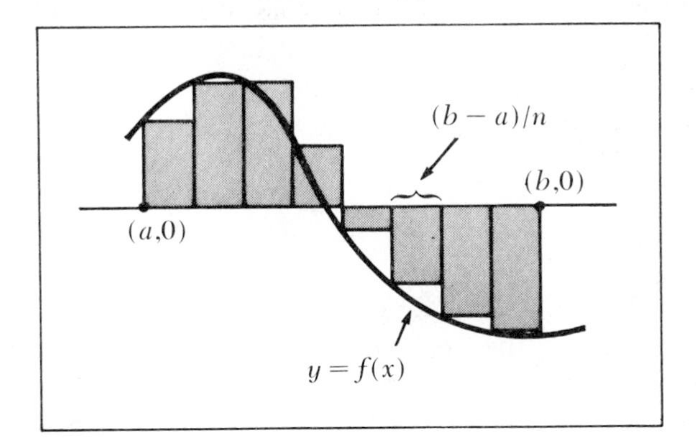

Figure 7–9 Expression (13) represents the sum of the areas of the shaded rectangles. This sum approximates the (signed) area bounded by the graph of f, the x-axis, and the lines $x = a$ and $x = b$, which is $\int_a^b f(x)\,dx$.

The integral is the sum of both positive and negative areas and may therefore have a value that contradicts our usual notion that area always has positive sign.

The portion of the area under the curve in the subinterval $[x_{i-1}, x_i]$ can, in most cases, be better approximated by means of a trapezoid (Figure 7–10) than by the rectangle with area $f(x_{i-1})(x_i - x_{i-1})$. For

TRAPEZOIDAL APPROXIMATION

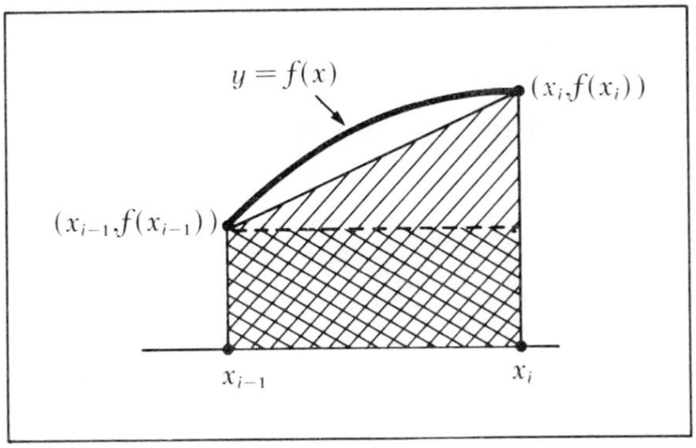

y = f(x)
(x_i,f(x_i))
(x_{i-1},f(x_{i-1}))
x_{i-1}
x_i

Figure 7–10 The portion of the area over the interval $[x_{i-1}, x_i]$ can usually be better approximated by means of a trapezoid than by means of a rectangle of the type whose areas are summed to obtain expression (13). The area of the rectangle is $f(x_{i-1})(x_i - x_{i-1})$; that of the trapezoid is $(1/2)(f(x_{i-1}) + f(x_i))(x_i - x_{i-1})$.

comparison, this rectangle is also indicated in Figure 7–10. Using the standard formula for the area of a trapezoid, we calculate the area in the figure to be

$$\tfrac{1}{2}(f(x_i) + f(x_{i-1}))(x_i - x_{i-1}) = \tfrac{1}{2}(f(x_i) + f(x_{i-1}))((b - a)/n)$$

$$= \frac{b - a}{2n}(f(x_i) + f(x_{i-1})) \tag{15}$$

when the interval $[a, b]$ is divided into n equal subintervals. This process is exactly analogous to the approximation of the area under the curve by means of rectangles, and we can sum the areas of the trapezoids in the same way that we summed rectangles to obtain sum (13). Letting $x_0 = a$, we note that, as before, $x_1 = a + (b - a)/n$, $x_2 = a + 2(b - a)/n$, and so on until $x_n = a + n(b - a)/n = b$. Substituting these values of x_i into expression (15) and adding together the areas of all the trapezoids, we obtain

$$\frac{b - a}{2n}\Bigg[\left(f(a) + f\left(a + \frac{b - a}{n}\right)\right) + \left(f\left(a + 2\frac{b - a}{n}\right) + f\left(a + \frac{b - a}{n}\right)\right) + \ldots + \left(f\left(a + \frac{(n - 1)(b - a)}{n}\right) + f\left(a + \frac{(n - 2)(b - a)}{n}\right)\right) + \left(f\left(a + \frac{n(b - a)}{n}\right) + f\left(a + \frac{(n - 1)(b - a)}{n}\right)\right)\Bigg].$$

Note that, with the exception of $f(a)$ and $f(b)$ $\left(\text{which is shown here as } f\left(a + \frac{n(b - a)}{n}\right)\right)$, each term in this sum appears twice. Thus we can simplify the sum to

$$\frac{b - a}{2n}\left[f(a) + 2\left(f\left(a + \frac{(b - a)}{n}\right)\right) + \ldots + 2\left(f\left(a + \frac{(n - 1)(b - a)}{n}\right)\right) + f(b)\right], \tag{16}$$

which is known as the *Trapezoidal Rule* for approximating $\int_a^b f(x)\,dx$. It generally provides a better approximation than does sum (13). As before, the approximation becomes more accurate with increasing values of n.

Note that the Trapezoidal Rule can be derived in other ways. For instance, we may consider the straight line between the points $(x_{i-1}, f(x_{i-1}))$ and $(x_i, f(x_i))$ to be an approximation to the graph of f between the lines $x = x_{i-1}$ and $x = x_i$ (cf. Figure 7–10). On the other hand, we may substitute the average value of $f(x_{i-1})$ and $f(x_i)$—namely $\frac{1}{2}(f(x_{i-1}) + f(x_i))$—for the term $f(x_{i-1})$ in sum (13). The rationale behind this substitution is that the average value describes the values that f assumes on the interval $[x_{i-1}, x_i]$ better than does $f(x_{i-1})$.

Example 11: Consider once more the integral

$$\int_1^2 \frac{e^x\,dx}{x}$$

of Examples 8 and 10. For $n = 4$ the endpoints of the four equal intervals are

$$1,\ 5/4,\ 3/2,\ 7/4,\ 2.$$

According to the Trapezoidal Rule, when $n = 4$, $\int_1^2 (e^x\,dx/x)$ is approximately

$$\frac{1}{8}\left[e + \frac{2e^{5/4}}{5/4} + \frac{2e^{3/2}}{3/2} + \frac{2e^{7/4}}{7/4} + \frac{e^2}{2}\right]. \tag{17}$$

The area represented by sum (17) appears in Figure 7–11. As was mentioned in Example 10, this sum can be evaluated using logarithms, a table of values of e^x, or a computer. It will be found that sum (14) is equal to 2.94, while sum (17) is equal to 3.06. This confirms the assertion that the trapezoid includes more of the area under the curve than does the rectangle, as shown in Figure 7–10.

Applying the Trapezoidal Rule is essentially a matter of substituting the appropriate values in sum (16). While the computations involved in evaluating parts or all of (16) are frequently very cumbersome, particularly for large values of n, such calculations are now often facilitated by computers. These observations also apply for the next method for approximating definite integrals, *Simpson's Rule.*

The Trapezoidal Rule entails the use of straight line segments to approximate the graph of f, which in turn helps us to approximate $\int_a^b f(x)\,dx$. Simpson's Rule, however, involves approximating the graph of f using arcs of parabolas. Again $[a, b]$ is subdivided into n equal

APPROXIMATING $\int_1^2 (e^x/x)\,dx$ BY THE TRAPEZOIDAL RULE

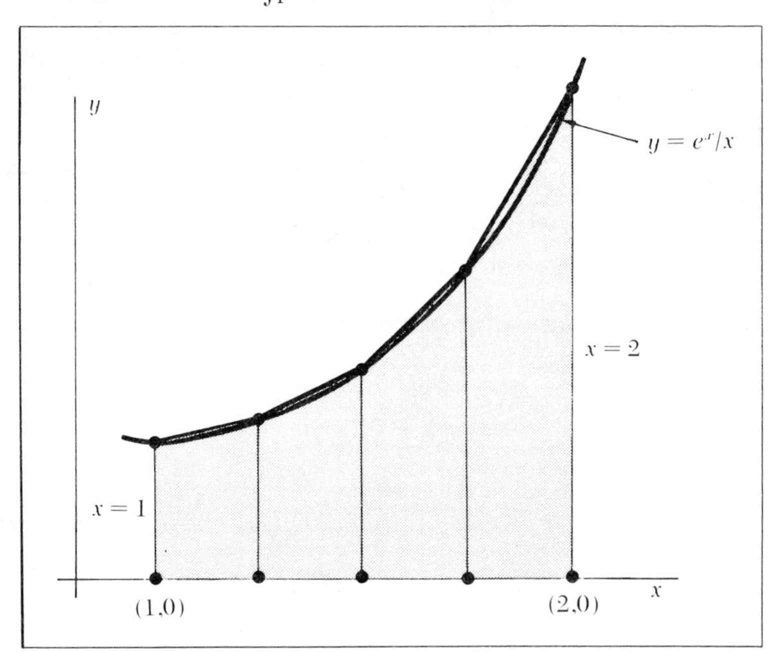

Figure 7–11 The approximation given by the Trapezoidal Rule with $n = 4$ is the sum of the areas of the trapezoids shown.

subintervals, but now n must be an *even* integer. The approximation of the graph of f by parabolas used in this method is illustrated in Figure 7–12. Again we let the end points of the subintervals be

$$x_0 = a, x_1, x_2, \ldots, x_{n-1}, x_n = b.$$

Simpson's Rule states:

If $\int_a^b f(x)\,dx$ exists, then it is approximated by

$$\left(\frac{b-a}{3n}\right)(f(a) + 4(f(x_1) + f(x_3)\ldots) + 2(f(x_2) + f(x_4) + \ldots) + f(b)). \tag{18}$$

We make no attempt here to derive Simpson's Rule.

Generally, although not always, Simpson's Rule produces a more accurate approximation than does the Trapezoidal Rule; moreover, the larger n is, the better the approximation. (We might expect this better approximation if we recall from Section 5.4 that a quadratic function,

SIMPSON'S RULE

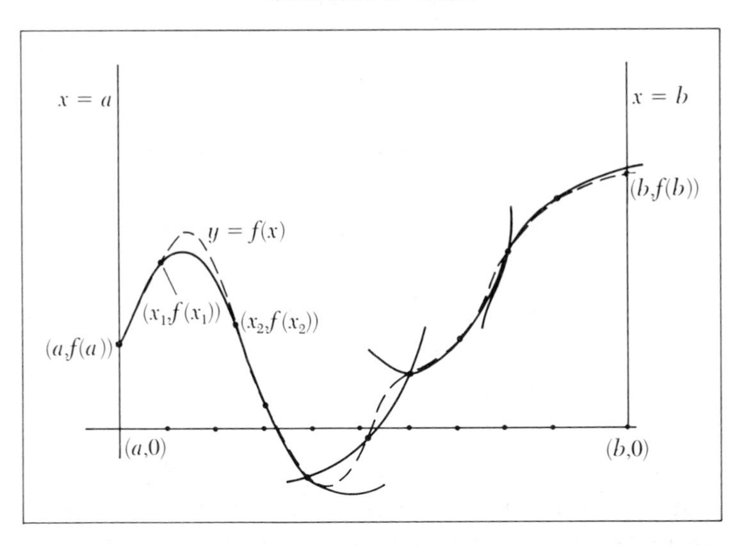

Figure 7–12 Simpson's Rule is derived by approximating the graph of the function f by a series of parabolas. Since a parabola usually gives a better approximation than a straight line, Simpson's Rule usually gives a better approximation to the area under the graph of f than does the Trapezoidal Rule for the same value of n.

whose graph is a parabola, usually yields a better approximation to a given function than does a linear function.)

Note that in expression (18) the sum of the function values is multiplied by 4 at the odd-numbered endpoints, and is multiplied by 2 at the even-numbered endpoints other than a or b.

As with the Trapezoidal Rule, the application of Simpson's Rule consists of choosing a value of n (remember that for Simpson's Rule n must be even), setting

$$x_i = a + i(b - a)/n, \qquad i = 0, 1, \ldots, n,$$

calculating the appropriate function values, and then substituting these in sum (18) to obtain the approximation to the definite integral.

Example 12: Approximate

$$\int_{-1}^{3} (1/\sqrt{2\pi})e^{-x^2/2}\, dx$$

using Simpson's Rule with $n = 4$. The endpoints of the subintervals are

$$-1, 0, 1, 2, 3.$$

The function value at the even-numbered endpoint other than -1 and 3 is

$$(1/\sqrt{2\pi})e^{-1/2},$$

while the function values at the odd-numbered endpoints are

$$(1/\sqrt{2\pi})(1) = 1/\sqrt{2\pi} \quad \text{and} \quad (1/\sqrt{2\pi})e^{-2}.$$

Substituting these values in sum (18) and factoring out $1/\sqrt{2\pi}$, we obtain

$$(1/3)(1)(1/\sqrt{2\pi})(e^{-1/2} + 4(1 + e^{-2}) + 2(e^{-1/2}) + e^{-9/2}), \tag{19}$$

which is approximately 0.845. The actual value of this integral, correct to four decimal places, is 0.8400; thus we observe that the approximation obtained from Simpson's Rule is quite accurate even with a relatively small value of n.

Exercises

ROUTINE

1. Approximate each of the following integrals using expression (13) with $n = 4$. Sketch the area represented by each integral together with that area found by using (13). Compute the integral exactly, where possible, to determine the accuracy of the approximation.

a) $\int_0^2 x\,dx$

b) $\int_1^2 (1/x)\,dx$

c) $\int_0^1 (1/(1 + x^2))\,dx$

d) $\int_{-1}^1 e^x\,dx$

e) $\int_{-1}^{-2} x^3\,dx$

f) $\int_1^2 (x/(x^2 + 1))\,dx$

g) $\int_1^5 (x^2 + 1)^{1/2}\,dx$

h) $\int_{0.5}^{2.5} \frac{e^x}{x}\,dx$

i) $\int_0^4 e^{-x^2}\,dx$

j) $\int_{-1}^1 x^2e^{-x^2}\,dx$

k) $\int_4^5 (e^x/\ln x)\,dx$

l) $\int_1^2 (1 + \ln x)^{1/2}\,dx$

2. Approximate each of the integrals given in Exercise 1 using the Trapezoidal Rule with $n = 4$. Sketch the area represented by each integral together with that area found using the Trapezoidal Rule.

3. Approximate each of the integrals given in Exercise 1 using Simpson's Rule with $n = 4$. In those instances where the integral can be computed exactly, determine whether expression (13), Simpson's Rule, or the Trapezoidal Rule gives the best approximation.

Review of Chapter 7

We define $\lim_{x\to\infty} f(x) = L$ if, given any $p > 0$, we can find a number M such that $x > M$ implies $|f(x) - L| < p$. Similarly, we define $\lim_{x\to-\infty} f(x) = L$ if, given any $p > 0$, we can find a number M such that $x < M$ implies $|f(x) - L| < p$. If f is defined on (a, b) (where a can be $-\infty$ and b can be ∞), and if f is integrable on $[u, v]$ for each u and v such that $a < u < v < b$, then we define the *improper integral* of f from a to b as

$$\int_a^b f(x)\,dx = \lim_{\substack{u\to a\\ v\to b}} \int_u^v f(x)\,dx.$$

The notion of the improper integral extends the notion of the definite integral and allows us to use the techniques of integration to solve a broader range of problems than would be solvable using the definite integral alone.

If the definite integral $\int_a^b f(x)\,dx$ is known to exist, it can be approximated as follows: Subdivide the interval $[a, b]$ into n equal subintervals with endpoints

$$x_0 = a, x_1 = a + ((b - a)/n), \ldots, x_i = a + i((b - a)/n), \ldots, x_n = b.$$

Then $\int_a^b f(x)\,dx$ can be approximated either by the Trapezoidal Rule,

$$\frac{b-a}{2n}[f(x_0) + 2f(x_1) + \ldots + 2f(x_{n-1}) + f(x_n)],$$

or by Simpson's Rule (in which n must be even),

$$\frac{b-a}{3n}[f(x_0) + 4(f(x_1) + f(x_3) + \ldots) + 2(f(x_2) + f(x_4) + \ldots) + f(x_n)].$$

REVIEW EXERCISES

1. Evaluate each of the following limits and improper integrals. If a limit does not exist or an integral fails to converge, write *diverges*.

a) $\lim_{x \to \infty} 1/(x + 1)$

b) $\lim_{x \to \infty} 1/(x + 3)$

c) $\lim_{x \to -\infty} 3/(7x - 8)$

d) $\lim_{x \to \infty} x/(x^2 + 1)$

e) $\int_0^\infty e^{-2x}\, dx$

f) $\int_0^\infty xe^{-x}\, dx$

g) $\int_0^1 (1/x^2)\, dx$

h) $\int_1^\infty (1/x^3)\, dx$

i) $\int_{-\infty}^\infty e^{-x}\, dx$

2. Approximate each of the following integrals using first the Trapezoidal Rule and then Simpson's rule, both with $n = 4$.

a) $\int_0^1 dx/(x^2 + 1)$

b) $\int_1^5 (1 + x^2)^{1/2}\, dx$

c) $\int_{-1}^1 e^{-x^2}\, dx$

d) $\int_0^4 \log_e (x^2 + 1)\, dx$

e) $\int_e^{e+1} e^x \log_e x\, dx$

f) $\int_0^1 \frac{x - 1}{x + 1}\, dx$

Vocabulary

function of *n* variables
partial derivative
constraint
Lagrangian multiplier
maxima and minima

Chapter 8

Functions of Several Variables

8.1 PARTIAL DIFFERENTIATION

In solving practical problems, one rarely encounters functions of just one variable. For instance, a person's income may depend on education, location, whom the person knows, and many other factors; in other words, income is a function of not just one variable, but of several. The volume of a crate does not depend solely on its height or its length, but simultaneously on three factors: length, height, and width. If x, y, and z denote the dimensions of the crate, then the volume V is designated by the function

$$V(x, y, z) = xyz. \tag{1}$$

The symbolic notation $V(x, y, z)$ indicates that V depends on all of the factors x, y, and z; once x, y, and z are specified, we can compute the volume V.

If a company buys x tons of metal at \$150 per ton, and y tons of sand at \$10 per ton, the total amount T that the company spends on both metal and sand is a function of both x and y:

$$T(x, y) = 150x + 10y. \tag{2}$$

The functions defined in equations (1) and (2) are examples of functions of several variables. In equation (1) the function V associates the real number xyz with the *ordered triple* (x, y, z), where x, y, and z are real numbers. In equation (2) the function T associates the real number $150x + 10y$ with an *ordered pair* (x, y) of real numbers.

Definition 1: *An expression of the form*

$$(x_1, x_2, \ldots, x_n), \tag{3}$$

where $x_1, \ldots, x_n$ are real numbers, is called an ***ordered n-tuple*** *of real numbers.*

A rule, formula, or expression that assigns a unique real number to each element of a set of ordered n-tuples of real numbers is said to be a ***function of n variables.*** *The set of n-tuples for which such a function is defined is said to be the* ***domain*** *of the function. The $x_1, \ldots, x_n$ are called the* ***function variables.***

Thus we see that the function defined in equation (1) is a function of the three variables x, y, and z, while the function defined in equation (2) is a function of the two variables x and y.

The graph of a function of one variable is a subset of the coordinate plane. The graphs of functions of more than one variable are subsets of higher dimensional spaces. For example, the graph of a function of two variables is a subset of 3-dimensional Euclidean space, the space of ordinary solid geometry. In general, the graph of a function of n variables is a subset of an $(n + 1)$*-dimensional space.*

Example 1: The graph of the function defined in equation (2) is a portion of a plane (Figure 8–1). Since the function variables cannot reasonably be negative—for in that case the company would be buying negative quantities of material—the domain of the function consists of all ordered pairs (x, y) such that x and y are non-negative.

We will not attempt to present the principles of graphing functions of more than one variable, nor will we discuss the structure of Euclidean n-dimensional space. Indeed, the study of multivariate functions is substantially more complicated than the study of functions of one variable, with which we have so far been concerned. In this chapter we will simply outline some of the basic computational techniques used in dealing with functions of several variables.

Again consider the function

$$V(x, y, z) = xyz.$$

If we assign fixed values to any two variables of this function, then the result is a function of a single variable. For example, setting $x = 1$ and

GRAPH OF A FUNCTION OF TWO VARIABLES

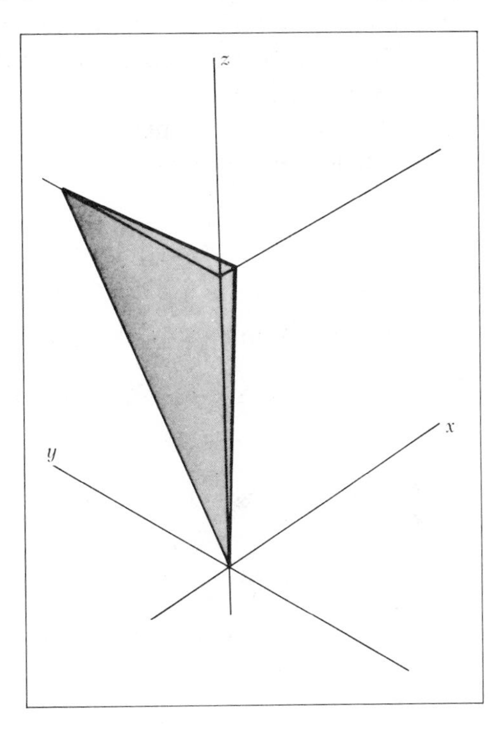

Figure 8–1 The graph of a function of two variables is a subset of three-dimensional Euclidean space. The graph of $T(x, y) = 150x + 10y$, $x \geq 0, y \geq 0$, for instance, is a portion of a plane.

$y = 2$, we obtain a function with z as its sole variable:

$$V(1, 2, z) = (1)(2)z = 2z. \tag{4}$$

It is possible to compute the first derivative of the function in equation (4) with respect to z by the standard methods discussed earlier. This derivative is

$$V'(1, 2, z) = 2. \tag{5}$$

We can interpret equation (5) as the slope of V when $x = 1$ and $y = 2$, or as the rate of change of V with respect to z when $x = 1$ and $y = 2$.

Example 2: Let the function S of two variables be defined by

$$S(x, y) = x^2 + y^2. \tag{6}$$

For any fixed value of x, $S(x, y)$ becomes a function of y alone; hence if we consider x to be constant, we can compute the derivatives of S with respect to y. This derivative is

$$S'_y(x, y) = 2y \tag{7}$$

We use the subscript y to indicate that we are considering $S(x, y)$ as a function of y alone and are computing the derivative with respect to y. Note that since x is considered a constant, the derivative of x^2 is 0; only y is considered as a variable in the computation of S'_y. We can interpret $S'_y(x, y)$ as the rate of change of S with respect to y for any fixed value of x.

If we let x equal some fixed value a, then

$$S(a, y) = a^2 + y^2, \tag{8}$$

which is the equation of a parabola. Thus we conclude that for any fixed value of x, the graph of S will be a parabola (Figure 8–2). We can likewise

A CROSS-SECTION OF $S(x, y) = x^2 + y^2$

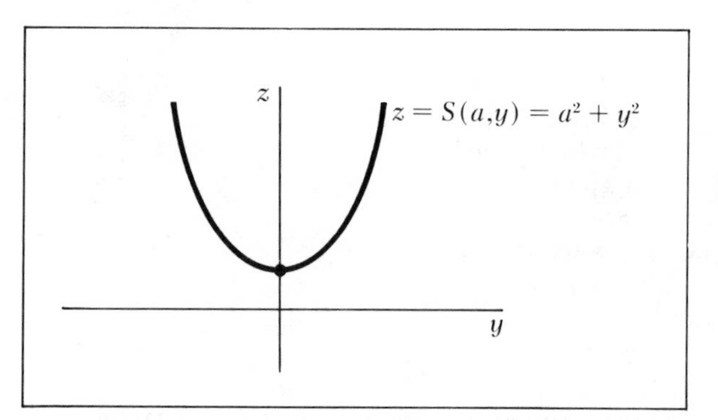

Figure 8–2 When $x = a$, the graph of $S(a, y) = a^2 + y^2$ is a parabola.

calculate $S'_x(x, y)$, the first derivative of S with respect to x, now considering y to be a constant:

$$S'_x(x, y) = 2x. \tag{9}$$

Note, too, that for any fixed value of y, the graph of S will again be a parabola.

Example 3: Let a function of three variables be defined by

$$f(x, y, z) = xy + e^{yz}. \tag{10}$$

In this case we can consider any two variables as constant and calculate the first derivative with respect to the remaining variable. Using notation analogous to that of Example 2, we obtain

$$f'_x(x, y, z) = y,$$
$$f'_y(x, y, z) = x + ze^{yz}, \tag{11}$$

and

$$f'_z(x, y, z) = ye^{yz}.$$

Confirm the accuracy of the derived equation (11).

Definition 2: *Suppose that f is a function of the variables $x_1, \ldots, x_n$. Then the first derivative with respect to x_i, where $i = 1, \ldots, n$, is called the* ***first partial derivative of f with respect to x_i****. We denote this first partial derivative of f with respect to x_i by either f_{x_i} or $\partial f/\partial x_i$.*

We compute f_{x_i} in the same way that we would compute the first derivative of f with respect to x_i if f were a function of x_i alone. In other words, we consider all of the variables of f to be constant except for x_i, and then we compute the first derivative in the usual manner. We see then that f_{x_i} is a function that measures how fast f is changing with respect to the variable x_i. If we use the notation of Definition 2, the derivatives in equations (11) are f_x, f_y, and f_z, or $\partial f/\partial x$, $\partial f/\partial y$, and $\partial f/\partial z$.

The partial derivative of a function of n variables is itself a function of n variables; hence we can compute its partial derivatives, which in turn are functions of n variables, and so on. We illustrate this point in the next example.

Example 4: Let $f(x, y, z) = x^2y^3z^4$. Then

$$f_x(x, y, z) = 2xy^3z^4. \tag{12}$$

We can compute the partial derivative of f_x with respect to any of its variables. For example,

$$(f_x)_y(x, y, z) = 6xy^2z^4, \tag{13}$$

while

$$(f_x)_x(x, y, z) = 2y^3z^4. \tag{14}$$

Usually the *second partial derivative* of f found in equation (13) would be denoted by either f_{xy} or $\partial^2 f/\partial x\, \partial y$, while that in equation (14) would be denoted by f_{xx} or $\partial^2 f/\partial x^2$.

We now define the successive partial derivatives.

Definition 3: *If f is a function of the variables $x_1, \ldots, x_n$, then each $(f_{x_i})_{x_j}$ is called a* **second partial derivative of f.** *We usually denote $(f_{x_i})_{x_j}$ by either $f_{x_i x_j}$ or $\partial^2 f/\partial x_i\, \partial x_j$.*

A first partial derivative of a second partial derivative is called a **third partial derivative,** *and so on.*

The notation for the higher partial derivatives follows the same pattern as that for the first and second partial derivatives. For example, the first derivative of $\partial^2 f/\partial x_1\, \partial x_2$ with respect to x_3 is denoted by

$$\partial^3 f/\partial x_1\, \partial x_2\, \partial x_3.$$

A function f of the variables x and y has four second partial derivatives: f_{xx}, f_{xy}, f_{yx}, and f_{yy}. In virtually all practical situations that one is likely to encounter, f_{xy} is equal to f_{yx}. We have $f_{xy} = f_{yx}$, for example, when both f_{xy} and f_{yx} are continuous.

Example 5: Let $f(x, y) = e^{x^2 y} - y$. Then the four second partial derivatives of f are given by

$$f_{xx}(x, y) = 2ye^{x^2 y} + 4x^2 y^2 e^{x^2 y},$$
$$f_{xy}(x, y) = f_{yx}(x, y) = 2xe^{x^2 y} + 2x^3 y e^{x^2 y},$$
$$f_{yy}(x, y) = x^4 e^{x^2 y}.$$

Here, as was the case with functions of one variable, the second derivative measures the rate of change of the first derivative, and thus, indirectly at least, provides information about the original function. In the next section we will explore one of the most important uses of the partial derivatives of a function of several variables.

Exercises

ROUTINE

1. Compute the first partial derivatives with respect to each variable of each of the functions defined below.

a) $f(x, y) = xy$
b) $f(x, y) = x^2 y + 3$
c) $g(x, y, z) = xy + yz + xz$
d) $T(x, y, z) = 10x + 65y - 17z$
e) $S(x, y) = (x^2 + y^2)^{1/2}$
f) $P(x, y, z) = \ln (xyz)$
g) $f(x, y, z) = e^{-xy^2} + \ln (x^2 z^3) + 5$
h) $f(x, y) = xy/(x + y)$
i) $f(x_1, x_2, x_3) = (x_1 + x_2^2 + x_3^3)^{16}$
j) $f(x, y) = x^y$
k) $g(w, z) = w^e e^z$
l) $h(x, y) = \log_x y$

2. Compute all of the first and second partial derivatives of each of the following functions.

a) $f(x, y) = xy^2$
b) $g(x, y) = xe^y$
c) $h(x, y) = x^2 - y^3$
d) $h(x, y, z) = x^2 + y^2 + z^2$
e) $E(x, y) = e^{xy}$

3. Sketch the graphs of the functions resulting in (a), (b), and (c) of Exercise 2 when:

a) $x = 0$
b) $y = 0$
c) $x = 1$
d) $y = 2$
e) $x = -1$

CHALLENGING

4. Find an expression that defines the function of several variables described in each of the following.

a) The total outside surface area of a box (including top) with dimensions x, y, and z.
b) The perimeter of a rectangular field with dimensions x and y.
c) The total cost of putting down a new floor with dimensions x and y, if tile costs $1.50 per square foot, baseboard costs $.50 per linear foot, and labor, which is required at the rate of one hour for every 20 square feet, costs $7.50 per hour.
d) The total profit realized from selling x items of type A and y items of type B, if the profit functions for A and B are

$$p_A(n) = n^2 - n + 1$$
$$p_B(n) = 45n/(16 - n^2).$$

8.2 MAXIMA AND MINIMA. LAGRANGIAN MULTIPLIERS

Example 6: Consider once again the function

$$S(x, y) = x^2 + y^2.$$

We have already observed in Example 2 that whenever x is assigned a constant value, the graph of the resulting function in y is a parabola. Likewise, when y is assigned a constant value, the graph of the resulting function in x is also a parabola. It is readily seen, either graphically or by using the techniques that we developed earlier for finding maxima and minima for functions of a single variable, that if $x = a$, then $S(a, y)$ has a minimum value when $y = 0$. Moreover, if $y = b$, then $S(x, b)$ has a minimum value at $x = 0$.

Thus if x is considered as a constant, $S(x, y)$ has a minimum value when $y = 0$; if y is considered as a constant, $S(x, y)$ has a minimum value when $x = 0$. Therefore, it is reasonable to suppose that $S(x, y)$

will have its minimum value when both x and y equal 0, for it is then that $S(x, y)$ will have a minimum value with respect to *both* x and y.

The maxima and minima with respect to either x or y (i.e., when one of the variables is assumed to be held constant) can be found using the methods involving the first derivative. For example, when x is assumed constant, we compute the first derivative with respect to y, locate the critical points of this function, and then test them to determine their nature. But the first derivative of S with respect to y is simply the first partial derivative of S with respect to y. The first partial derivatives can then be used to locate the relative maxima and minima of a function of several variables.

Definition 4: *The function f of the variables $x_1, \ldots, x_n$ is said to have a* ***relative maximum*** *(or a* ***relative minimum****) at the point $(a_1, \ldots, a_n)$ if an open interval U can be found that contains each of the a_i such that*

$$f(x_1, \ldots, x_n) \leq f(a_1, \ldots, a_n)$$

(or $f(x_1, \ldots, x_n) \geq f(a_1, \ldots, a_n)$ in the case of the relative minimum) provided that each x_i is in U.

We say that f has a ***critical point*** *at $(a_1, \ldots, a_n)$ if the value of each first partial derivative at $(a_1, \ldots, a_n)$ is equal to zero:*

$$f_{x_i}(a_1, \ldots, a_n) = 0 \qquad \text{for } i = 1, \ldots, n.$$

Thus, a function f of the two variables x and y has a critical point at (a, b) if $f_x(a, b) = 0$ *and* $f_y(a, b) = 0$.

Example 7: The function $S(x, y) = x^2 + y^2$ has a critical point and a relative minimum at $(0, 0)$; in fact, it attains its absolute minimum value of 0 at $(0, 0)$. If either x or y is different from 0, $S(x, y)$ will be greater than 0.

Example 8: Consider the function

$$T(x, y) = x^2 - y^2.$$

Then $T_x(x, y) = 2x$, while $T_y(x, y) = -2y$. If a point (a, b) is to be a critical point of T, then we must have $T_x(a, b) = 2a = 0$ and $T_y(a, b) = -2b = 0$ simultaneously. This can happen only if $a = b = 0$. Hence we see that $(0, 0)$ is a critical point (the only critical point) of T. We now show that T has neither a relative maximum nor a relative minimum at $(0, 0)$.

When $y = 0$,

$$T(x, y) = T(x, 0) = x^2 - 0^2 = x^2.$$

Therefore, when $y = 0$ the graph of T is a parabola which has a minimum at $x = 0$ (Figure 8–3). This, in turn, implies that if $(x, 0)$ is chosen close to but different from $(0, 0)$, then $T(x, 0) = x^2$ will be larger than $T(0, 0) = 0$. Therefore, $T(0, 0)$ is certainly not the largest value of $T(x, y)$ for (x, y) close to $(0, 0)$.

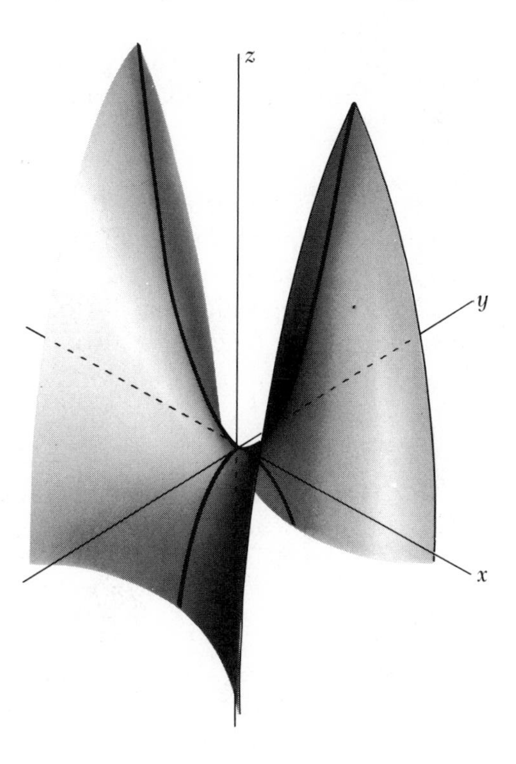

Figure 8–3 A three-dimensional representation of the graph of $T(x, y) = x^2 - y^2$. When $y = 0$, the graph of $T(x, 0)$ is a parabola with a minimum at $x = 0$. When $x = 0$, the graph of $T(0, y)$ is a parabola with a maximum at $y = 0$. Thus, T has neither a maximum nor a minimum at $(0, 0)$.

On the other hand, when $x = 0$, then

$$T(x, y) = T(0, y) = 0^2 - y^2 = -y^2.$$

Therefore, when $x = 0$ the graph of T is a parabola which has a maximum value at $y = 0$ (Figure 8–3). This, in turn, implies that if $(0, y)$ is chosen close to but different from $(0, 0)$, then $T(0, y)$ will be smaller than $T(0, 0) = 0$. Therefore $T(0, 0)$ is certainly not the smallest value of $T(x, y)$ for (x, y) close to $(0, 0)$.

Since $T(0, 0)$ is neither the largest nor the smallest function value of T for (x, y) close to $(0, 0)$, T has neither a relative maximum nor a relative minimum at $(0, 0)$.

How do we determine where a function of several variables has a relative maximum or minimum? This is a fairly difficult question to answer; partial answers appear in the following two propositions.

Proposition 1: If

a) f is a function of the variables $x_1, \ldots, x_n$,

b) f has a relative maximum or minimum at $(a_1, \ldots, a_n)$,

and

c) all of the first partial derivatives of f exist at $(a_1, \ldots, a_n)$,

then f has a critical point at $(a_1, \ldots, a_n)$.

Thus, if f is a function of the variables x and y that has a relative maximum or minimum at (a, b) and for which $f_x(a, b)$ and $f_y(a, b)$ exist, then $f_x(a, b) = f_y(a, b) = 0$.

The next proposition sets forth a method for testing the nature of a critical point for a function of two variables.

Proposition 2: Suppose that f is a function of the two variables x and y and that (a, b) is a critical point of f. Evaluate

$$\left(\frac{\partial^2 f}{\partial x^2}\right)\left(\frac{\partial^2 f}{\partial y^2}\right) - \left(\frac{\partial^2 f}{\partial x\, \partial y}\right) \tag{15}$$

at (a, b). If

a) expression (15) is positive and $\partial^2 f/\partial x^2 + \partial^2 f/\partial y^2$ is negative at (a, b), then f has a relative maximum at (a, b);

b) expression (15) is positive and $\partial^2 f/\partial x^2 + \partial^2 f/\partial y^2$ is positive at (a, b), then f has a relative minimum at (a, b);

c) expression (15) is negative, then f has neither a relative maximum nor a relative minimum at (a, b); and

d) expression (15) is equal to zero, then the test yields no information.

The use of Propositions 1 and 2 is illustrated in the next example.

Example 9: Let

$$f(x, y) = x^2 - xy + 2y^2.$$

Then

$$f_x(x, y) = 2x - y$$

and

$$f_y(x, y) = -x + 4y.$$

To find any critical points of f we must solve the equations

$$2x - y = 0$$
$$-x + 4y = 0$$

simultaneously. Doing so, we see that (0, 0) is the only critical point of f.

We now evaluate expression (15). Since $f_{xx}(0, 0) = 2$, $f_{yy}(0, 0) = 4$, and $f_{xy}(0, 0) = -1$, expression (15) in this case becomes $2 \cdot 4 + 1 = 9$, which is positive.

Having determined that (15) is positive, we see that parts (a) or (b) of Proposition 2 may enable us to determine the nature of the critical point (0, 0). We therefore evaluate $\partial^2 f/\partial x^2 + \partial^2 f/\partial y^2$ at (0, 0). This latter quantity is $2 + 4 = 6$, which is positive. Therefore, by part (b) of Proposition 2, f has a relative minimum at (0, 0).

In solving maximum-minimum problems of functions of only one variable, we saw that the physical context of the problem often limited the solutions we would accept. For example, the length of a box can never turn out to be negative. Thus, if we seek the length x of a certain kind of box in order to maximize the volume of the box, the solution is subject to the *constraint* that $x \geq 0$. Similarly, in maximum-minimum problems involving functions of several variables, the solutions will be required to satisfy certain constraints due either to the nature of the problem or to limitations that one imposes on the problem. We illustrate this concept in the following examples.

Example 10: The material for the sides of a carton costs $.35 per square foot, while the material for the top and bottom of the box costs $.50 per square foot. What are the dimensions of the least expensive carton that has a volume of 1.5 cubic feet?

We let the carton's dimensions be x, y, and z (see Figure 8–4). Then the cost $C(x, y, z)$ of the carton in terms of x, y, and z is

$$\begin{aligned} C(x, y, z) &= (0.35)(2xz + 2yz) + (0.50)(2xy) \\ &= (0.7)(xz + yz) + xy. \end{aligned} \tag{16}$$

A MINIMIZATION PROBLEM IN THREE VARIABLES

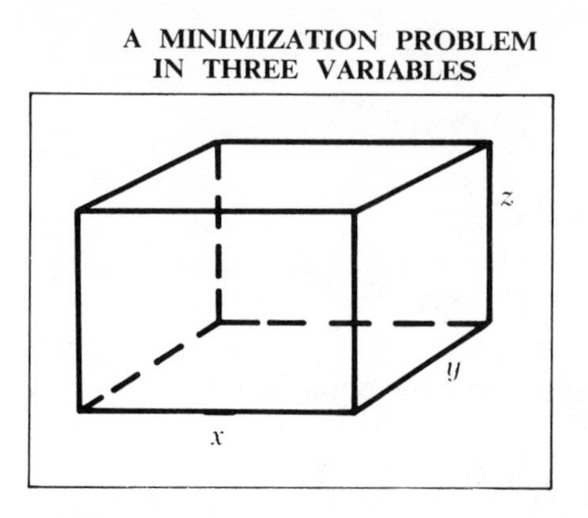

Figure 8–4 A carton is to have dimensions x, y, and z. We must find values for the variables such that the carton will have a volume of 1.5 cubic feet and the cost of the carton will be a minimum.

The object is to minimize $C(x, y, z)$. The variables x, y, and z must be non-negative since they represent lengths. However, we do not simply search for x, y, and z values that make $C(x, y, z)$ a minimum. These values must also satisfy the added condition concerning the volume of the carton:

$$V(x, y, z) = xyz = 1.5. \tag{17}$$

One method of solution is to solve equation (17) for z in terms of x and y and then substitute the result into equation (16). Thus, equation (17) becomes

$$z = (1.5)/xy,$$

and substituting this value into equation (16), we obtain

$$C(x, y) = (1.05)\left(\frac{1}{y} + \frac{1}{x}\right) + xy. \tag{18}$$

We can then locate any relative minima of equation (18) using the methods outlined in Propositions 1 and 2.

A technique which is sometimes useful in finding relative maxima and minima involves the use of *Lagrangian multipliers*. This method is widely used when the function to be maximized or minimized is subject to some constraint. A *constraint* is a condition which we require the variables of a function to satisfy; for example, we may demand that they all be positive, or that none of them be less than -3. We will confine our discussion to the use of Lagrangian multipliers in solving problems, since a discussion of why this technique works is beyond the scope of this text.

Let f be a function of the variables $x_1, \ldots, x_n$ subject to some constraint. We shall require that the constraint be expressible as an equation of the form

$$g(x_1, \ldots, x_n) = 0, \tag{19}$$

where g is an appropriate function. For example, the constraint that the variables add up to 1 can be written

$$x_1 + x_2 + \ldots + x_n - 1 = 0.$$

We define a new function h with variables $x_1, \ldots, x_n$ and a new variable:

$$h(x_1, \ldots, x_n, \lambda) = f(x_1, \ldots, x_n) - \lambda g(x_1, \ldots, x_n). \tag{20}$$

We call λ a *Lagrangian multiplier.*

Example 11: Set $f(x, y) = x^2 - y^2$ and let f be subject to the constraint $2x - 3y - 4 = 0$. Then the function h in this particular case will be

$$h(x, y, \lambda) = x^2 - y^2 - \lambda(2x - 3y - 4).$$

The following statement sums up the basic theorem about Lagrangian multipliers:

Proposition 3: If f is a function of the variables $x_1, \ldots, x_n$ that is defined and has all of its first partial derivatives defined for all values of $x_1, \ldots, x_n$ and which is subject to the constraint

$$g(x_1, \ldots, x_n) = 0,$$

then if the function h as defined in equation (20) has a critical point and a relative maximum or minimum at $(a_1, \ldots, a_n, \lambda_0)$, then f itself will have a critical point at $(a_1, \ldots, a_n)$ at which f will also have a relative maximum or minimum subject to the given constraint.

Thus, if f is a function of the variables $x_1, \ldots, x_n$ subject to some constraint, we follow these steps to apply the method of Lagrangian multipliers to find candidates for maxima and minima of f subject to the given constraint.

1. Express the constraint in the form (19). If this cannot be done, then the method of Lagrangian multipliers cannot be used.
2. Verify that f is defined and has all first partial derivatives defined for all values of $x_1, \ldots, x_n$.
3. Form the function h defined in equation (20).

4. Find the critical points of h. You will have to solve the equations

$$h_{x_1}(x_1, \ldots, x_n, \lambda) = 0$$

$$\ldots$$

$$h_{x_n}(x_1, \ldots, x_n, \lambda) = 0$$

$$h_\lambda(x_1, \ldots, x_n, \lambda) = 0$$

simultaneously.

A solution $(a_1, \ldots, a_n, \lambda_0)$ of this system of equations gives a relative minimum or maximum $(a_1, \ldots, a_n)$ of f which is also subject to the given constraint if h itself has a relative maximum or minimum at $(a_1, \ldots, a_n, \lambda_0)$. Unfortunately, the technique of Lagrangian multipliers does not give us the exact nature of the critical point $(a_1, \ldots, a_n)$, although this is often clear from the context of the problem.

We illustrate the method of Lagrangian multipliers by solving the problem posed in Example 10.

Example 12: The function to be minimized is expressed in equation (16); the constraint, which appears in equation (17), can also be written as $xyz - 1.5 = 0$. Therefore, $g(x, y, z) = xyz - 1.5$ expresses the constraint in the form (19). In this particular instance, the function h of equation (20) becomes

$$\begin{aligned} h(x, y, z, \lambda) &= C(x, y, z) - \lambda g(x, y, z) \\ &= (0.7)(xz + yz) + xy - \lambda(xyz - 1.5) \qquad (21) \\ &= 0.7xz + 0.7yz + xy - \lambda xyz + \lambda 1.5. \end{aligned}$$

We now determine the critical points of h. First note that

$$h_x(x, y, z, \lambda) = 0.7z + y - \lambda yz, \qquad (22)$$

$$h_y(x, y, z, \lambda) = 0.7z + x - \lambda xz, \qquad (23)$$

$$h_z(x, y, z, \lambda) = 0.7x + 0.7y - \lambda xy, \qquad (24)$$

and

$$h_\lambda(x, y, z, \lambda) = -xyz - 1.5. \qquad (25)$$

To find the critical points of h, we must set each of the partial derivatives equal to zero and simultaneously solve the resulting equations,

$$0.7z + y - \lambda yz = 0, \qquad (26)$$

$$0.7z + x - \lambda xz = 0, \qquad (27)$$

$$0.7x + 0.7y - \lambda xy = 0, \qquad (28)$$

and

$$-xyz - 1.5 = 0. \qquad (29)$$

If we subtract equation (27) from equation (26) and simplify the result, we obtain

$$(y - x)(1 - \lambda z) = 0,$$

which yields

$$y = x \quad \text{and} \quad \lambda z = 1. \tag{30}$$

Multiplying equation (26) by x, multiplying equation (28) by z, and subtracting one from the other yields

$$xy - 0.7yz = 0$$

or

$$x = 0.7z.$$

As we found above, $x = y$; thus we can write

$$x = y = 0.7z.$$

We can use the condition that $xyz = 1.5$ to solve directly for x, y, and z:

$$0.7x^3 = 1.5,$$

or

$$x = (1.5/0.7)^{1/3},$$

$$y = (1.5/0.7)^{1/3},$$

and

$$z = 0.7(1.5/0.7)^{1/3}.$$

This is the solution to the problem.

We conclude this section with another example illustrating the use of Lagrangian multipliers.

Example 13: What is the largest rectangular area that can be enclosed with exactly 100 feet of fencing? Let x and y be the sides of the rectangular area. Then the area itself is

$$A(x, y) = xy. \tag{31}$$

We want to maximize $A(x, y)$, but we are constrained by having only 100 feet of fencing. The perimeter of the rectangle must therefore be 100. We can express this as

$$2x + 2y - 100 = 0. \tag{32}$$

We then set

$$h(x, y, \lambda) = xy - \lambda(2x + 2y - 100).$$

The critical points of h are found by simultaneously solving the equations

$$h_x(x, y, \lambda) = y - 2\lambda = 0, \tag{33}$$

$$h_y(x, y, \lambda) = x - 2\lambda = 0, \tag{34}$$

and

$$h_\lambda(x, y, \lambda) = 2x + 2y - 100. \tag{35}$$

Subtracting equation (34) from equation (33), we obtain

$$y - x = 0,$$

from which we conclude that $x = y$. We then substitute y for x in equation (35) to obtain $4x = 100$, or $x = y = 25$. Thus, the rectangle is a square with side 25.

Note that in this case the functions involved are simple enough that equation (32) can be solved for y in terms of x, the result substituted into equation (31), and the new function of one variable differentiated to find the critical points of A. The real value of the method of Lagrangian multipliers appears when the function and constraints are complicated and difficult to differentiate.

Exercises

ROUTINE

1. Find the critical points of each of the following functions.
   a) $f(x, y) = xy^2$
   b) $g(x, y) = x^2 + xy + y^2$
   c) $f(x, y) = xe^y - y^2$
   d) $f(x, y) = x^2 - 3x + 2y + y^2$
   e) $f(x, y, z) = xyz + (xy/z)$
   f) $f(x, y) = (x + y)/xy$
   g) $g(x, y, z) = e^{xyz} + e^{-xyz} + z^2$
   h) $h(x, y, z) = (x^2 - y^2)/((x^2 + y^2)z)$
   i) $h(x, y, z) = \ln(x + y + z + 3) - \ln(xyz)$
   j) $f(x, y) = (x + y)^3(x - 4)^2$

2. Using Proposition 2 of this chapter, test the nature of each of the critical points found for functions (a) to (d) in Exercise 1.

3. Show that the rectangle of maximum area that can be enclosed with a fence of length L is a square with side $L/4$; in other words, generalize the result of Example 13. The same format used to solve that example can be used here, with L substituted for 100.

4. Solve Example 13 in the following manner. First, use equation (32) to obtain an expression for y in terms of x. Next, substitute this expression into equation (31) to obtain a function of the single variable x. Finally, compute the maximum value of this function. In this instance, the method of Lagrangian multipliers yields exactly the same result as does the method of reworking equation (31) to take the restraint into account.

5. Find three numbers whose sum is 25 and the sum of whose squares is a maximum. [Hint: Let the numbers be x, y, and z. Then maximize $x^2 + y^2 + z^2$, subject to the constraint $x + y + z = 25$.]

6. Find three numbers whose product is a maximum and whose sum is 10.

CHALLENGING

7. The walls of a room cost \$.70 per square foot; the floor costs \$1.00 per square foot; and the ceiling costs \$.80 per square foot (doors and windows are figured in the same way as regular parts of the wall). What are the dimensions of the largest room that can be built for \$5000?

8. Using the same costs as in Exercise 7, compute the dimensions of the cheapest room that can be built with a volume of 8000 cubic feet.

9. On the line whose equation is $y = 4x - 1$, find the point closest to the origin (0, 0). See Figure 8–5.

10. Find the relative minima of equation (18).

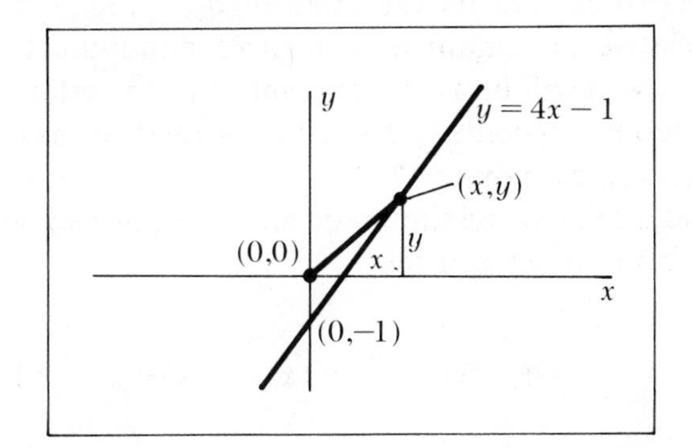

Figure 8–5 The situation in Exercise 9. The distance from (0, 0) to (x, y) is $\sqrt{x^2 + y^2}$.

Review of Chapter 8

A *function f of n variables* $x_1, \ldots, x_n$ is a rule, phrase, or relationship assigning a unique real number $f(x_1, \ldots, x_n)$ to each ordered n-tuple $(x_1, \ldots, x_n)$ of real numbers for which the function is defined. Let f be such a function.

By considering all except x_1 of $x_1, \ldots, x_n$ to be constant, we can treat f as if it were a function of the single variable x_1. We can thus compute the first derivative $\partial f/\partial x_1$ of f with respect to x_1; $\partial f/\partial x_1$ is called the *first partial derivative of f with respect to* x_1. In a similar manner we can calculate the first partial derivatives of f with respect to each variable. We can calculate higher partial derivatives by computing partial derivatives of the partial derivatives.

A point $(a_1, \ldots, a_n)$ at which all of the first partial derivatives of f are equal to 0 is called a *critical point*. If f has first partial derivatives with respect to each of its variables at $(a_1, \ldots, a_n)$, and $f(a_1, \ldots, a_n)$ is either a relative maximum or a relative minimum function value of f, then $(a_1, \ldots, a_n)$ will be a critical point of f. In certain instances there are tests to determine whether f has either a relative maximum or minimum at any of its critical points.

To find a relative maximum or minimum of f subject to the *constraint* $g(x_1, \ldots, x_n) = 0$, we can set

$$h(x_1, \ldots, x_n, \lambda) = f(x_1, \ldots, x_n) - \lambda g(x_1, \ldots, x_n).$$

If the critical point $(a_1, \ldots, a_n, \lambda_0)$ of h is a relative maximum or minimum of h, then $(a_1, \ldots, a_n)$ will bear the same relationship to f subject to the given constraint. This technique of solving relative maxima and minima subject to constraints is called the *method of Lagrangian multipliers.*

REVIEW EXERCISES

1. (i) Calculate all first partial derivatives of the following functions. (ii) Locate all critical points of these functions. (iii) Using Proposition 2 or any other reasonable means, try to identify the nature of each critical point.

a) $f(x, y) = x^2 - y$
b) $f(x, y) = e^x - e^{-xy}$
c) $f(x, y) = \log_e x + \log_e y - xy$
d) $f(x, y, z) = x + y + z - xyz$
e) $f(x, y) = xy/(x^2 + y^2)$
f) $f(x, y, z) = xe^y + ye^x - z/x - z/y$
g) $f(x, y) = (x + y)^4(x - y)^5$
h) $f(x, y, z) = x^2 + y^2 + z^2 - xyz$

2. a) In (a) of Exercise 1, determine all critical points of the function, subject to the constraint $x + y = 10$.
b) In (d) of Exercise 1, indicate all critical points of the function, subject to the constraint $3x - 2y + 4z = 10$.
c) Try to identify the nature of the critical points found in parts (a) and (b) above.

3. Find three numbers x, y, and z whose product is 18 and whose sum is a maximum.

4. A blend of three coffees A, B, and C is to be produced. Coffees A, B, and C cost \$.30, \$.45, and \$.50 per pound. The producer assigns flavor factors of 0.6, 0.8, and 0.9 to A, B, and C and estimates the flavor factor of the mixture by

$$0.6x^2 + 0.8y^2 + 0.9z^2,$$

where x, y, and z are the fractions per pound of A, B, and C in the mixture. What is the maximum flavor factor that can be obtained if the mixture is to be produced for exactly \$.43 per pound?

APPENDIX

Indefinite Integrals (Antiderivatives)

$$\int u^n\,du = \frac{1}{n+1}\,u^{n+1}$$

$$\int \frac{du}{u} = \ln u$$

$$\int a^u\,du = \frac{a^u}{\ln a}$$

$$\int e^u\,du = e^u$$

$$\int u^m e^{au}\,du = \frac{u^m e^{au}}{a} - \frac{m}{a}\int u^{m-1}e^{au}\,du \qquad (n > 0)$$

$$\int \frac{du}{u^2+a^2} = \frac{1}{a}\tan^{-1}\frac{u}{a} = -\frac{1}{a}\cot^{-1}\frac{u}{a}$$

$$\int \frac{du}{u^2-a^2} = \frac{1}{2a}\ln\left(\frac{u-a}{u+a}\right) = -\frac{1}{a}\coth^{-1}\frac{u}{a} \qquad (u^2 > a^2)$$

$$\int \frac{du}{a^2-u^2} = \frac{1}{2a}\ln\left(\frac{a+u}{a-u}\right) = -\frac{1}{a}\tanh^{-1}\frac{u}{a} \qquad (u^2 < a^2)$$

$$\int \frac{du}{\sqrt{a^2-u^2}} = \sin^{-1}\frac{u}{a} = -\cos^{-1}\frac{u}{a}$$

$$\int \frac{du}{\sqrt{u^2\pm a^2}} = \ln\left(u+\sqrt{u^2\pm a^2}\right)$$

$$\int \frac{du}{\sqrt{u^2+a^2}} = \sinh^{-1}\frac{u}{a}$$

$$\int \frac{du}{\sqrt{u^2-a^2}} = \cosh^{-1}\frac{u}{a}$$

$$\int \cos u\,du = \sin u$$

$$\int \sin u\,du = -\cos u$$

$$\int \sec^2 u\,du = \tan u$$

$$\int \operatorname{cosec}^2 u\,du = -\cot u$$

$$\int \tan u\,du = -\ln\cos u$$

$$\int \cot u\,du = \ln\sin u$$

$$\int \sinh u\,du = \cosh u$$

$$\int \cosh u\,du = \sinh u$$

$$\int \tanh u\,du = \ln\cosh u$$

$$\int \coth u\,du = \ln\sinh u$$

$$\int \sin^n u\,du = -\frac{1}{n}\sin^{n-1}u\cos u + \frac{n-1}{n}\int \sin^{n-2}u\,du$$

$$\int \cos^n u\,du = \frac{1}{n}\cos^{n-1}u\sin u + \frac{n-1}{n}\int \cos^{n-2}u\,du$$

APPENDIX

Definite Integrals

$$\int_0^{\pi/2} \sin^n u\, du = \int_0^{\pi/2} \cos^n u\, du$$

$$= \frac{1\cdot 3\cdot 5\cdot \cdots \cdot (n-1)}{2\cdot 4\cdot 6\cdot \cdots \cdot n}\cdot\frac{\pi}{2} \qquad (n \text{ even})$$

$$= \frac{2\cdot 4\cdot 6\cdot \cdots \cdot (n-1)}{1\cdot 3\cdot 5\cdot \cdots \cdot n} \qquad (n \text{ odd})$$

$$\int_0^{2\pi} \sin u\, du = \int_0^{2\pi} \cos u\, du = 0$$

$$\int_0^{2\pi} \sin^2 u\, du = \int_0^{2\pi} \cos^2 u\, du = \pi$$

$$\int_0^{2\pi} \sin u \cos u\, du = 0$$

$$\int_0^{\pi} \sin^2 m\,u\, du = \int_0^{\pi} \cos^2 m\,u\, du = \frac{\pi}{2}$$

$$\int_0^{\pi} \sin k\,u \cdot \sin m\,u\, du = \int_0^{\pi} \cos k\,u \cos m\,u\, du = 0 \qquad k \neq m$$

$$\int_0^{\infty} \frac{\sin m\,u\, du}{u} = \frac{\pi}{2} \qquad m > 0$$

$$= 0 \qquad m = 0$$

$$= -\frac{\pi}{2} \qquad m < 0$$

$$\int_0^{\infty} \frac{\cos u\, du}{u} = \infty$$

$$\int_0^{\infty} \frac{\tan u\, du}{u} = \frac{\pi}{2}$$

$$\int_0^{\infty} \frac{a\, du}{a^2 + u^2} = \frac{\pi}{2} \qquad a > 0$$

$$= 0 \qquad a = 0$$

$$= -\frac{\pi}{2} \qquad a < 0$$

$$\int_0^{\infty} \frac{\sin^2 u}{u^2}\, du = \frac{\pi}{2}$$

$$\int_0^{\infty} \frac{\cos m\,u\, du}{1 + u^2} = \frac{\pi}{2} e^{-m} \qquad m > 0$$

$$= \frac{\pi}{2} e^{m} \qquad m < 0$$

$$\int_0^{\infty} e^{-ax}\, dx = \frac{1}{a} \qquad a > 0$$

$$\int_0^{\infty} e^{-a^2u^2}\, du = \frac{\sqrt{\pi}}{2a}$$

$$\int_0^{\infty} u\, e^{-u^2}\, du = \frac{1}{2}$$

$$\int_0^{\infty} u^2 e^{-u^2}\, du = \frac{\sqrt{\pi}}{4}$$

$$\int_0^{\infty} e^{-au} \cos m\,u\, du = \frac{a}{a^2 + m^2} \qquad a > 0$$

$$\int_0^{\infty} e^{-au} \sin m\,u\, du = \frac{m}{a^2 + m^2} \qquad a > 0$$

Answers to Selected Exercises

CHAPTER 1

Section 1.1

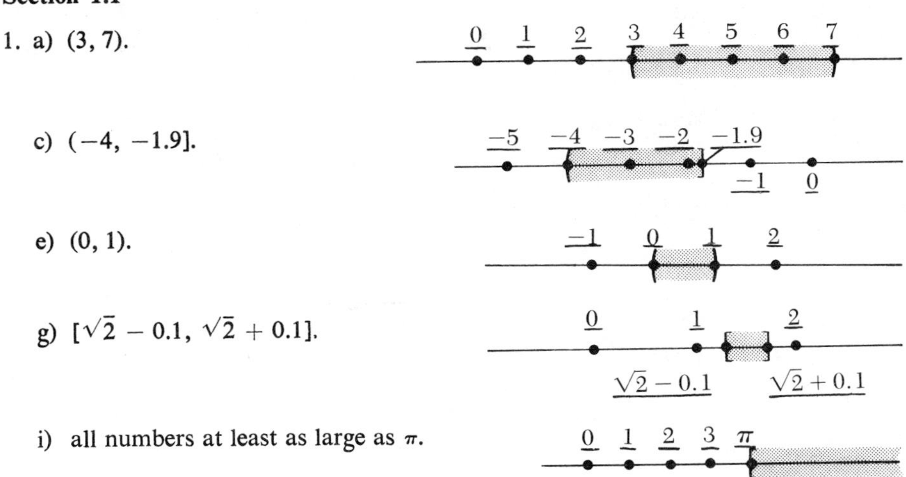

1. a) $(3, 7)$.

   c) $(-4, -1.9]$.

   e) $(0, 1)$.

   g) $[\sqrt{2} - 0.1, \sqrt{2} + 0.1]$.

   i) all numbers at least as large as π.

   k) the set containing no numbers at all.

2. a) union: $\{x \mid x$ is either red or a book$\}$; intersection: $\{x \mid x$ is a red book$\}$.
   c) union: all integers; intersection: the set containing no elements.

3. $a - n$; $b - p$; $c - e$; $d - i$; $f - o$; $g - l$; $h - j$; $k - m$.

4. a) $x^2 = 1$ if and only if $x = 1$ or $x = -1$.
   c) x is both brown and a book if and only if x is a brown book.

5. a) If S is any set, then since any element of S is an element of S, S is a subset of itself.
   c) If x is an element of $S \cap T$, then x is an element of both S and T. Therefore, $S \cap T$ is a subset of S and is also a subset of T.

6. Since $m^2/n^2 = 2$, then $m^2 = 2n^2$; hence 2 divides m^2. If m is odd, then $m = 2k + 1$ for some integer k, and therefore $m^2 = 4k^2 + 4k + 1$, which is not divisible by 2. Therefore, m must be even, or $m = 2k$ for some integer k. But then $m^2 = 4k^2 = 2n^2$, or $2k^2 = n^2$. Since 2 divides n^2, then by the same argument n itself must be even. Consequently, 2 divides both m and n, contradicting the assumption that m and n are not both divisible by 2.

8. a) $A \cup B = A$ since $B \subset A$; $A \cap C =$ set of brick dormitories; $B \cup D = D$ since $B \subset D$; $C \cap D =$ set of brick buildings that contain beds.

Section 1.2

1. Always true: a, b, c, d, i, j, k, m; sometimes true: e, f, g, l; never true: h.

2. a) $x < 9$; c) $x < -3/2$; e) $x < 2$; g) $(4, 22/3)$; i) $[-7, 13]$; k) $(-3/4, 0)$; m) $\{x \mid x < 1/4 \text{ or } x > 1/2\}$; o) $(-1, 0) \cup \{x \mid x > 1\}$.

3. a) $[1/3, 3/2)$; c) $(1, 5/4]$; e) $(3, 63)$.

4. a) $x + y < 8$; c) $x^2 + y^2 < (1/5)(x + y)$.

Section 1.3

1. a) all real numbers, $f(2) = 2, f(0) = 0$; c) all real numbers, $h(2) = 7, h(0) = -1$; e) $\{t \mid t \geq 1\}$, $m(2) = 1$, $m(0)$ not defined; g) all real numbers, $f(2) = -10$, $f(0) = -8$; j) $\{x \mid x \neq 0\}, g(2) = 1/4$.

2. a) functions a, c, d, e, f, g, h.

3.

(−1,5)
(1,1)
(−1/2,1/4)
(√2,0)
(−1,−5)
(1,−5)

4. a) $z = (y + 1)/2$; c) $z = y - 8$; e) $z = y^2$.

5. a)

6.

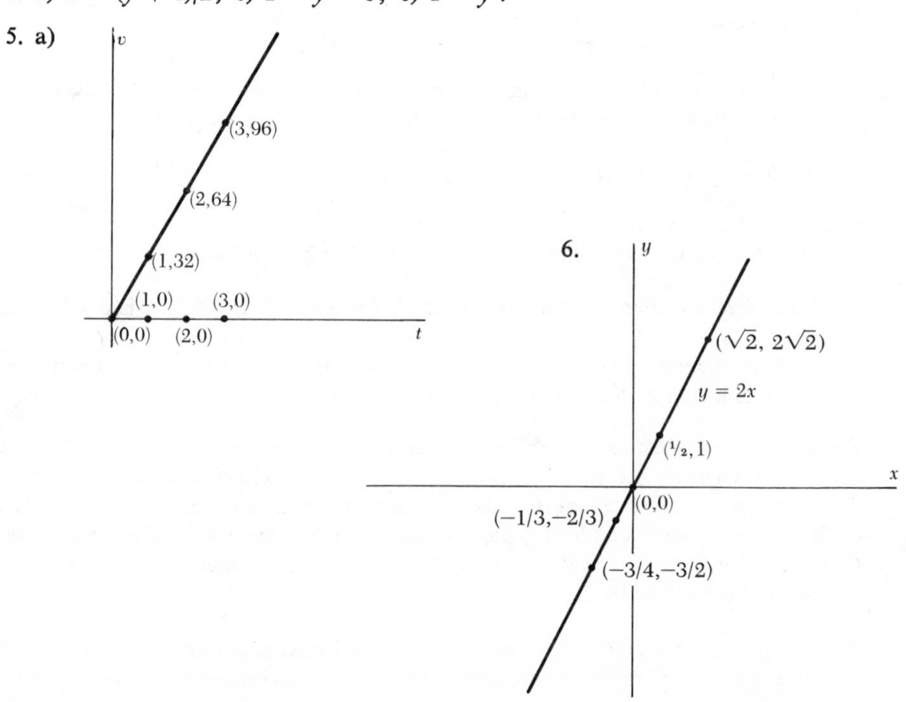

Section 1.4

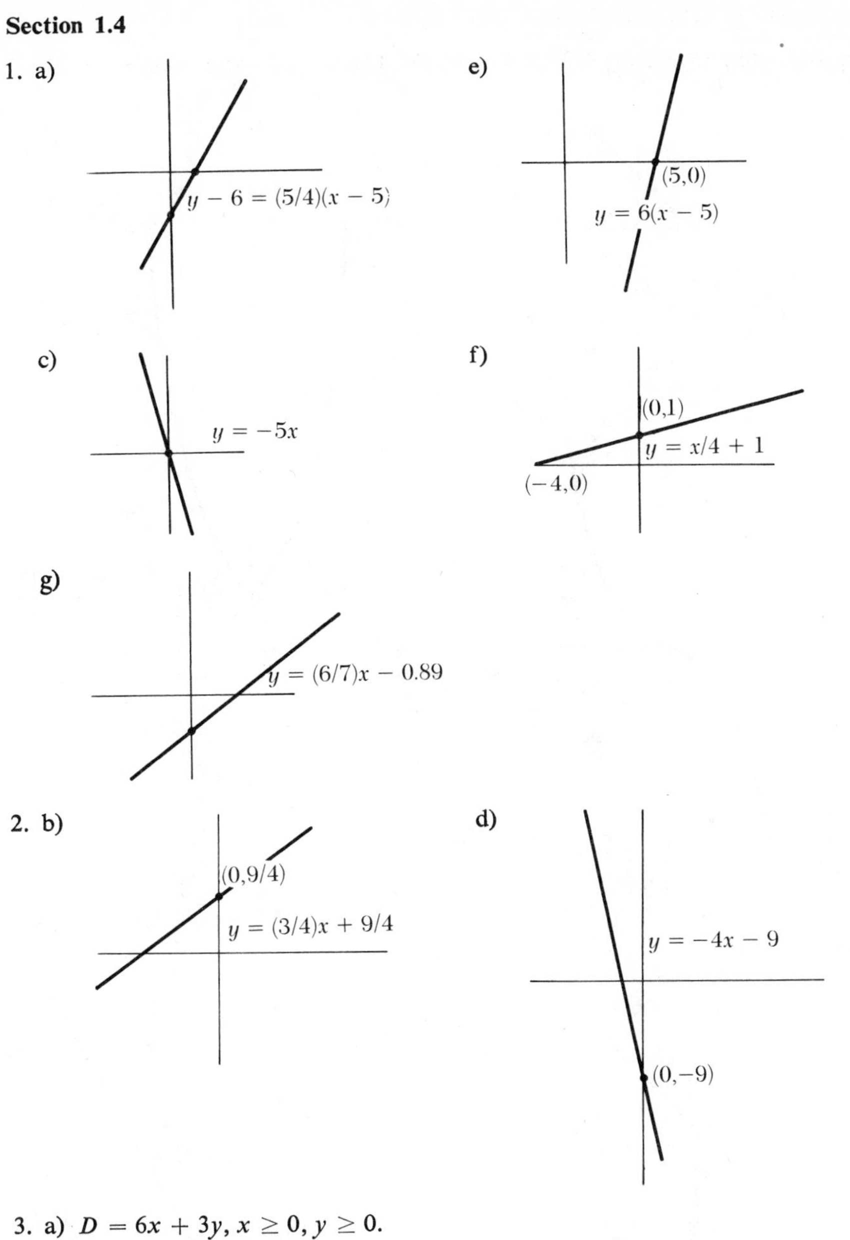

3. a) $D = 6x + 3y$, $x \geq 0$, $y \geq 0$.
   b) For $D = 10$, (a) gives $10 = 6x + 3y$, $x \geq 0$, $y \geq 0$.

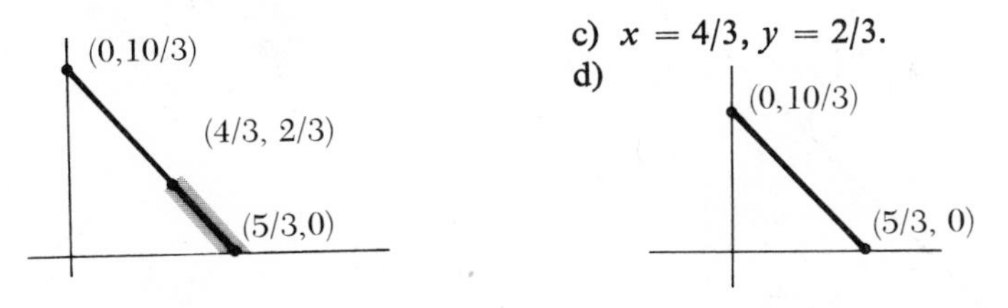

   c) $x = 4/3$, $y = 2/3$.
   d)

4. a) (3, 11); b) (1/2, 1/2); c) (3, −5).

5. approximately 1578 using (23, 2000) and (26, 1840) to compute slope $= -53\frac{1}{3}$.

6. a) d)

c)

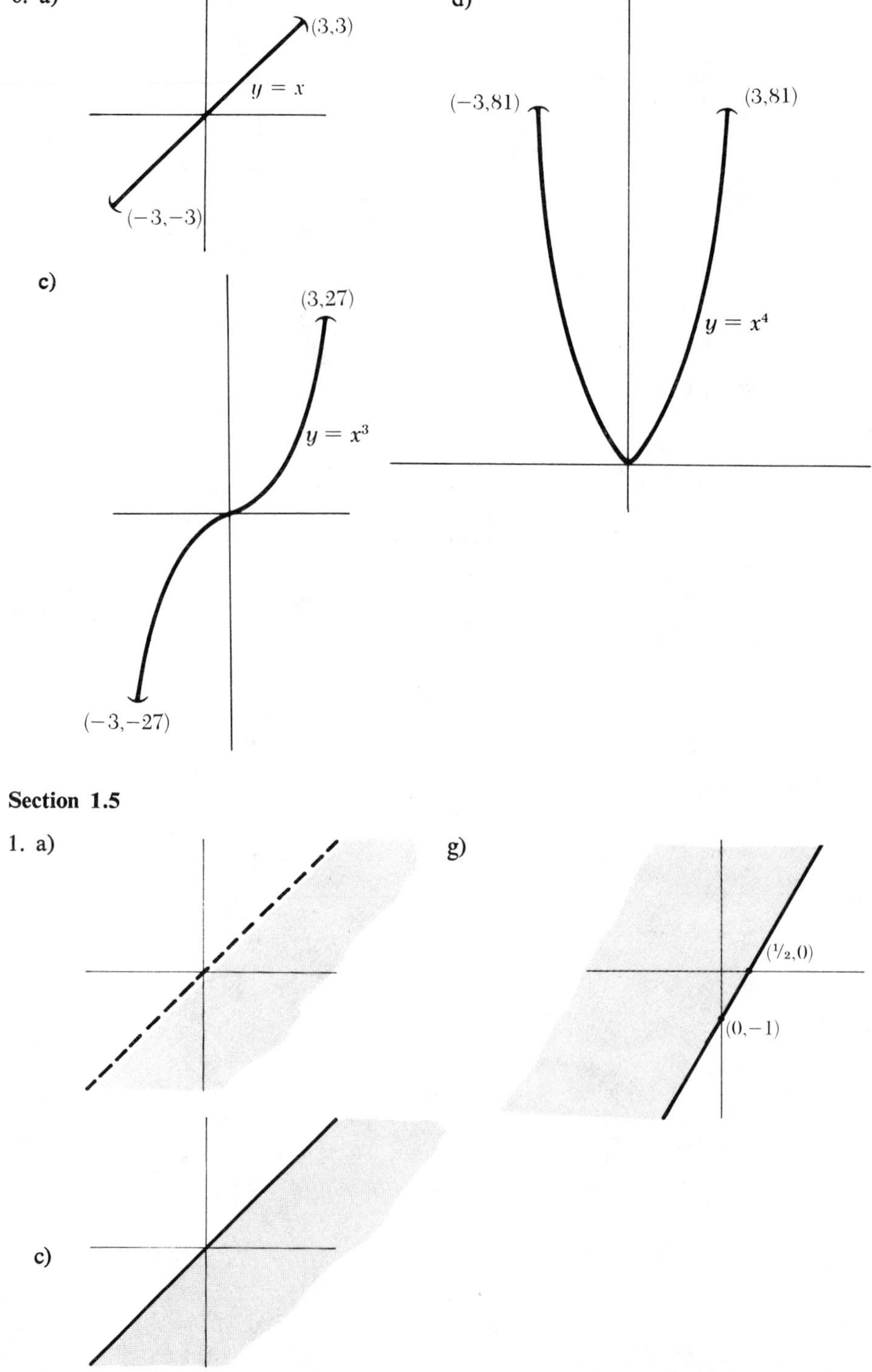

Section 1.5

1. a) g)

c)

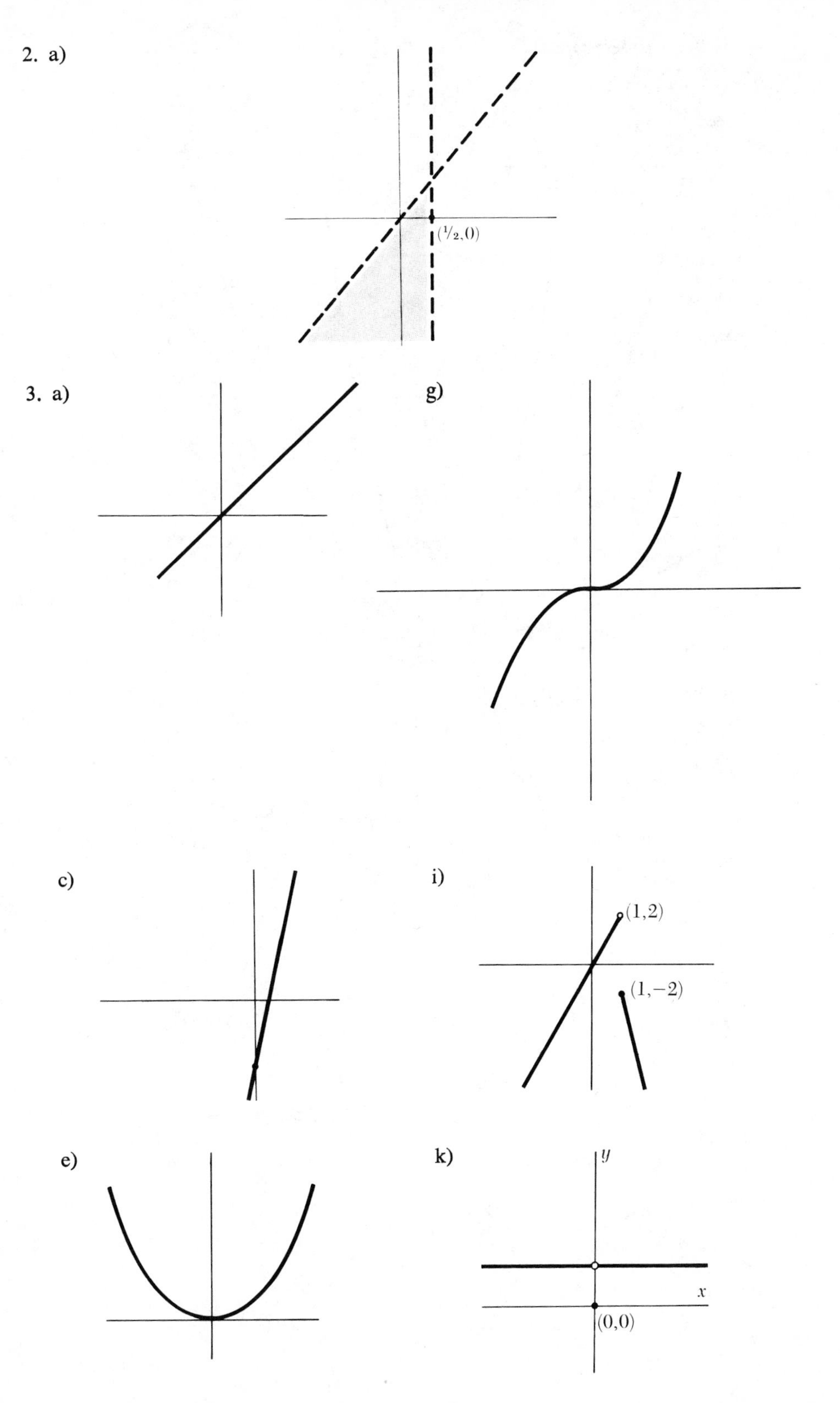
2. a)
(1/2,0)
3. a)
g)
c)
i)
(1,2)
(1,−2)
e)
k)
y
x
(0,0)

4. a) $A(x) = x(50 - x)$.
5. 4 miles per hour per minute.
7. 3/5 feet per minute.
8. a)

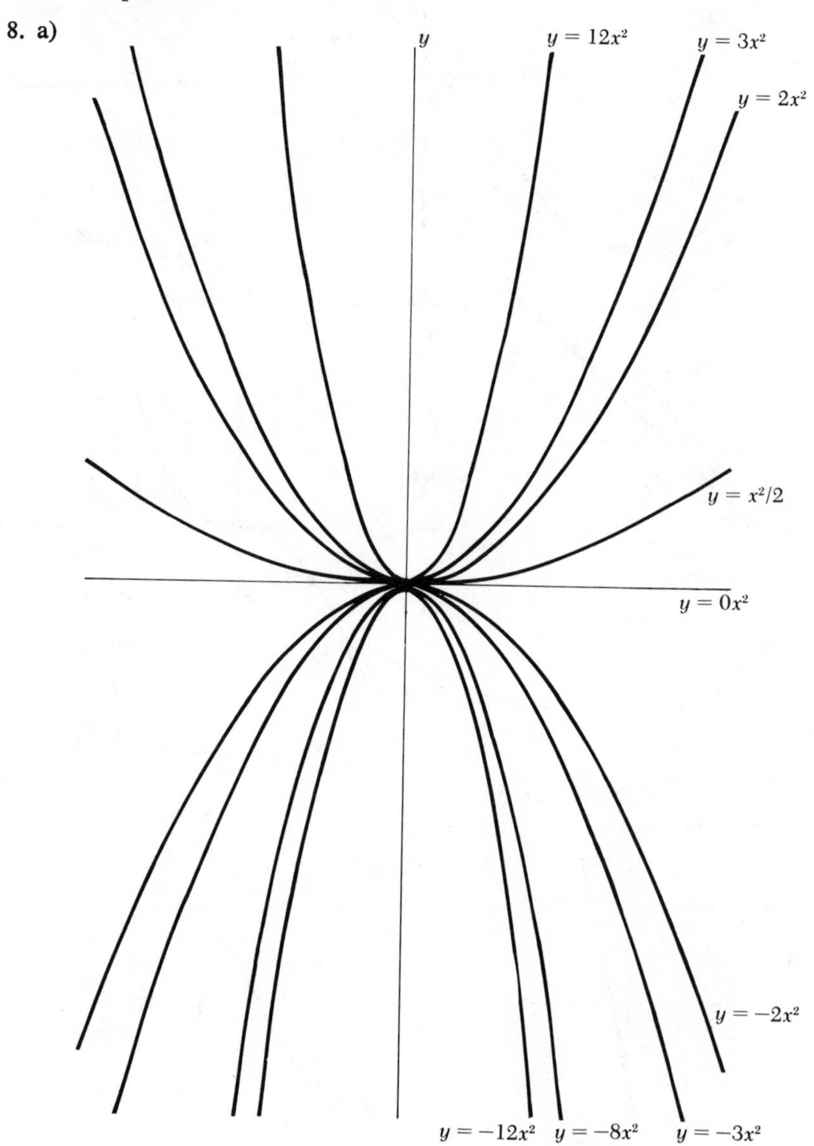

c)

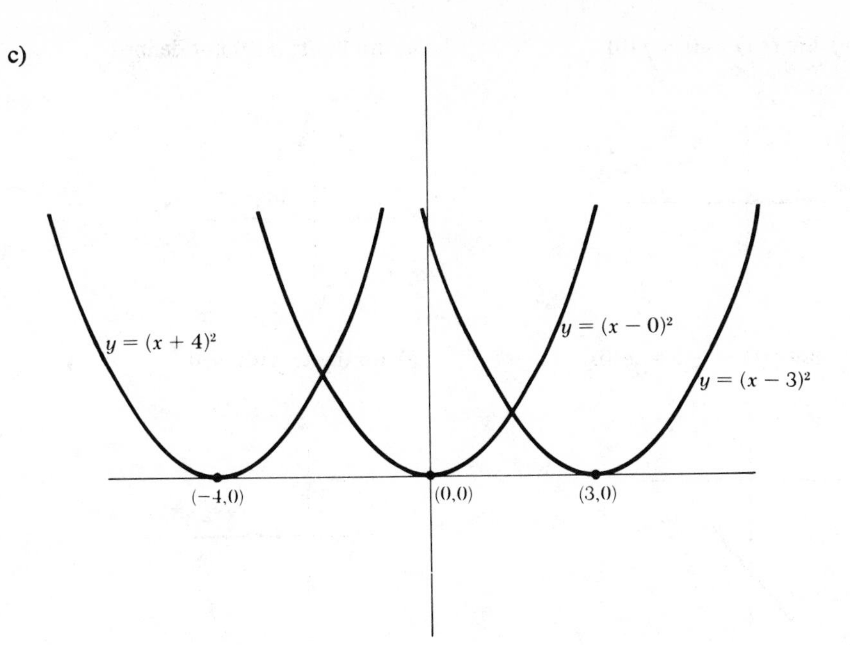

CHAPTER 2

Section 2.1

1. a) 1, 1, 1, 1, 1

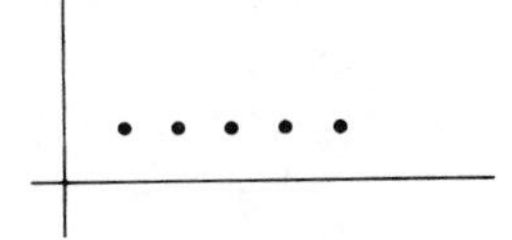

c) 1/2, 1/3, 1/4, 1/5, 1/6

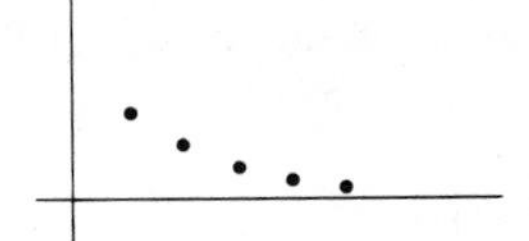

e) 1, 2, 3, 4, 5

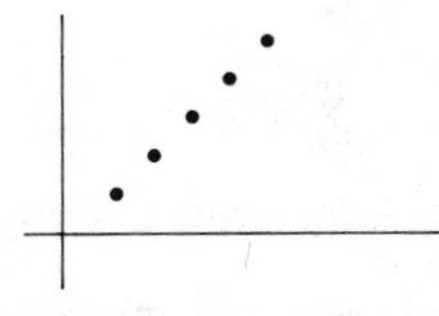

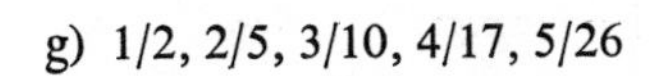

g) 1/2, 2/5, 3/10, 4/17, 5/26

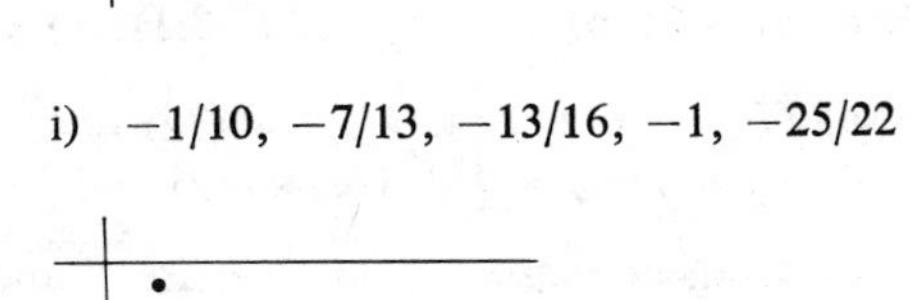

i) −1/10, −7/13, −13/16, −1, −25/22

2. a) $\lim_{x\to 0} f(x) = 0 = f(0)$

e) no limit; $h(0)$ not defined

c) $\lim_{t\to 0} g(t) = -5 = g(0)$

g) no limit; $f(0) = 0$

j) $g(0)$ not defined, no limit

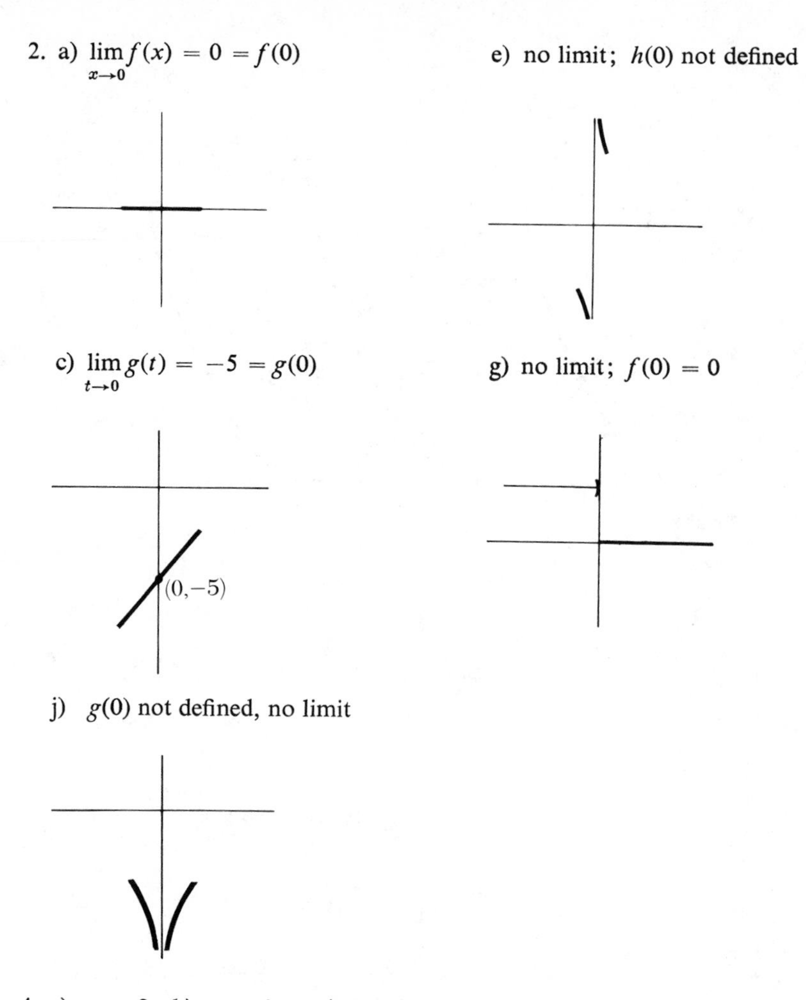

4. a) $s_n = 2$; b) $s_n = 2 + 1/(n + 1)$; e) $s_1 = 9$, $s_2 = 28$, $s_n = 9$, $n \geq 3$.

6. a) $f(s_n) = f(2) = 4$, $f(s_n) \to f(2) = 4$; c) $f(s_n) = f(\sqrt{4 + 1/n}) = (\sqrt{4 + 1/n})^2 = 4 + 1/n$, $f(s_n) = 4 + 1/n \to f(2) = 4$.

7. Convergent sequences in Exercise 1 and their limits: a) 1; c) 0; d) 0; g) 0; h) 3/5; i) -2; all others *no limit*.

8. a) 10; c) -4; e) $2b$; f) 3; g) 4; i) -3.

Section 2.2

1. a) continuous for all real numbers;
   c) continuous for all real numbers, sum of $-9x^2$ and -2;
   e) discontinuous only at $x = 3$, $f(3)$ not defined;
   g) continuous for all real numbers, $f(x) = x$ or $-x$, and $\lim_{x\to 0} f(x) = 0 = f(0)$;
   i) continuous for all real numbers, $g(x) = x$ or $2x$, $x \neq 0$, $\lim_{x\to 0} g(x) = 0 = g(0)$;
   k) discontinuous only at $t = 0$, $\lim_{t\to 0} m(t) = 0 \neq m(0) = 3$. We can make m continuous at 0 by redefining $m(0)$ to be 0.

2. a) c) e) g)

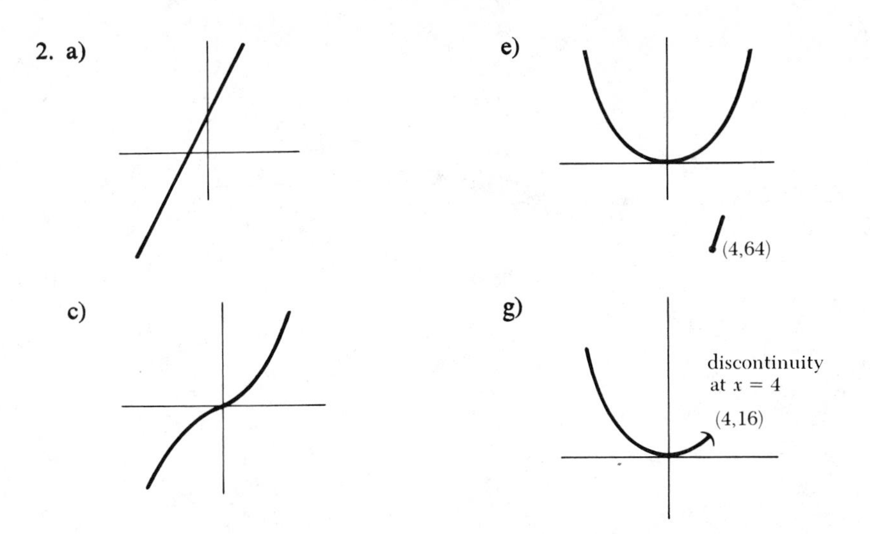

3. continuous: b, c, d; discontinuous: (a) takes only integer values; (e) likely to jump rather than to change continuously, and values are expressible only to two decimal places.

4. a) $(3/n) + 1$; b) 0, 1; c) 1, as $s_n \to 0, f(s_n) \to f(0)$ since $\lim_{x \to 0} f(x) = f(0) = 1$.

5. Answers in the following table.

(i)	(iii)	(v)
a) $s_n + t_n = 3$ $s_n t_n = 2$	3, 3, 3, 3, 3 2, 2, 2, 2, 2	3 2
b) $s_n + t_n = 1$ $s_n t_n = (1/n)(1 - 1/n)$	1, 1, 1, 1, 1 0, 1/4, 2/9, 3/16, 4/25	1 0
c) $s_n + t_n = n(1 + (-1)^n)$ $s_n t_n = (-1)^n n^2$	0, 1, 0, 3, 0 $-1, 4, -9, 16, -25$	no limit no limit
d) $s_n + t_n = \dfrac{3n}{(4n - 1)} - \dfrac{6n}{(7 - 5n)}$ $s_n t_n = -18n^2/((4n - 1)(7 - 5n))$	-2, 102/21, 270/88, . . . -3, 72/21, 162/88, . . .	39/20 18/20

6. a) positive for $x > 0$, negative for $x < 0$; c) positive for $x < 2$ and $x > 3$, negative for $2 < x < 3$; e) positive for $-7 < x < 3$ and $5 < x$, negative for $x < -7$ and $3 < x < 5$; h) positive for $x > 1$, negative for $x < 1$.

7. a) $f(a)$; b) a; c) $f(a)$.

CHAPTER 3

Section 3.1

1. a) 1/3; b) 8; c) $1/\dot{3}2$; d) $1/10^8 = 0.00000001$; f) 1/2; h) $1/10^9 = 0.000000001$.

2. a)

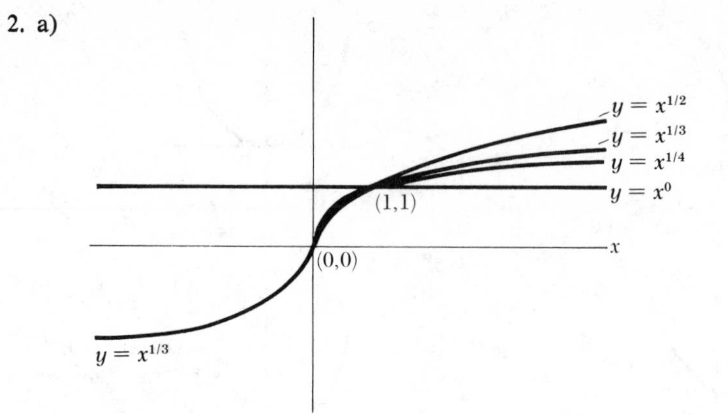
$y = x^{1/2}$
$y = x^{1/3}$
$y = x^{1/4}$
$y = x^0$
(1,1)
x
(0,0)
$y = x^{1/3}$

d)

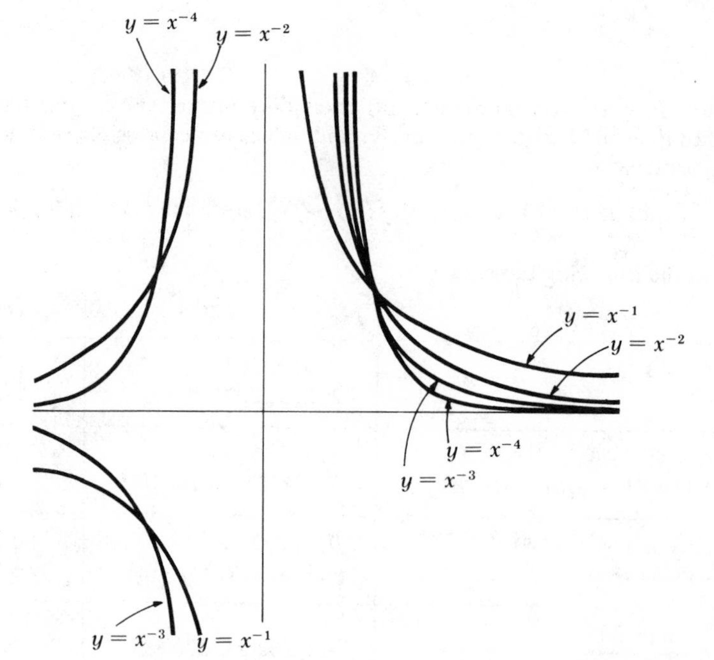
$y = x^{-4}$
$y = x^{-2}$
$y = x^{-1}$
$y = x^{-2}$
$y = x^{-4}$
$y = x^{-3}$
$y = x^{-3}$
$y = x^{-1}$

e)

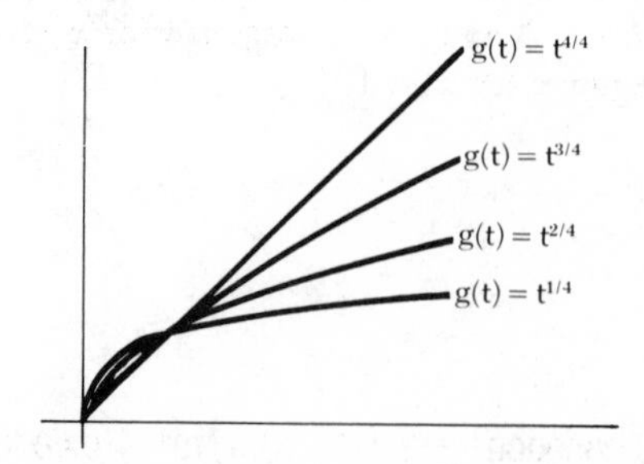
$g(t) = t^{4/4}$
$g(t) = t^{3/4}$
$g(t) = t^{2/4}$
$g(t) = t^{1/4}$

3. When $n = 25$, the value is $1000 + 100(5) = 1500$.

4. a)

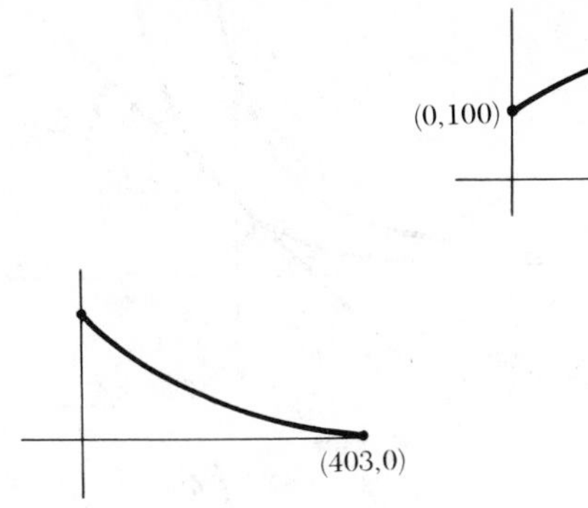

b) \$403.

5. $a^n b^n = (a \cdot a \cdot a \cdot \ldots n \text{ times})(b \cdot b \cdot b \cdot \ldots n \text{ times}) = (ab)(ab) \ldots n \text{ times}) = (ab)^n$

Section 3.2

1. a) 1.2040; c) −1.2552; e) −0.3010; g) 3.5562; i) 1/0.4343; k) 0.1793; m) −0.9542/0.4343; o) −0.3090; q) −0.9445.

2. a) g) c) i) e) k)

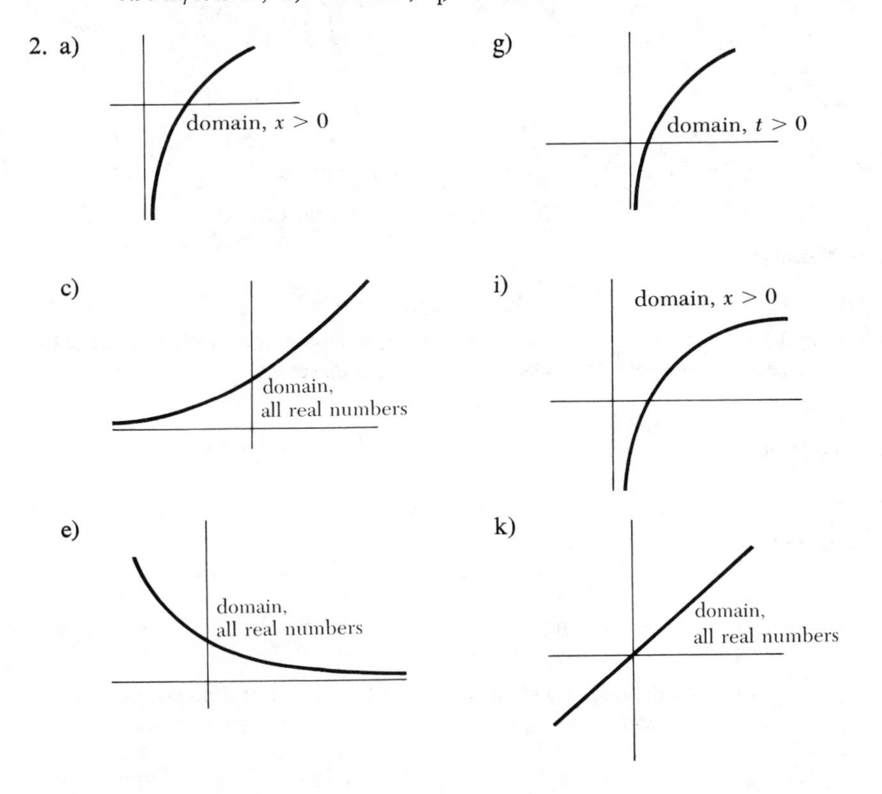

4. a) $a^t b^{-t} = a^t (b^{-1})^t = (ab^{-1})^t = (a/b)^t$.

5. bacterial growth, economic growth, population growth.

Section 3.3

1. a) e) c) g)

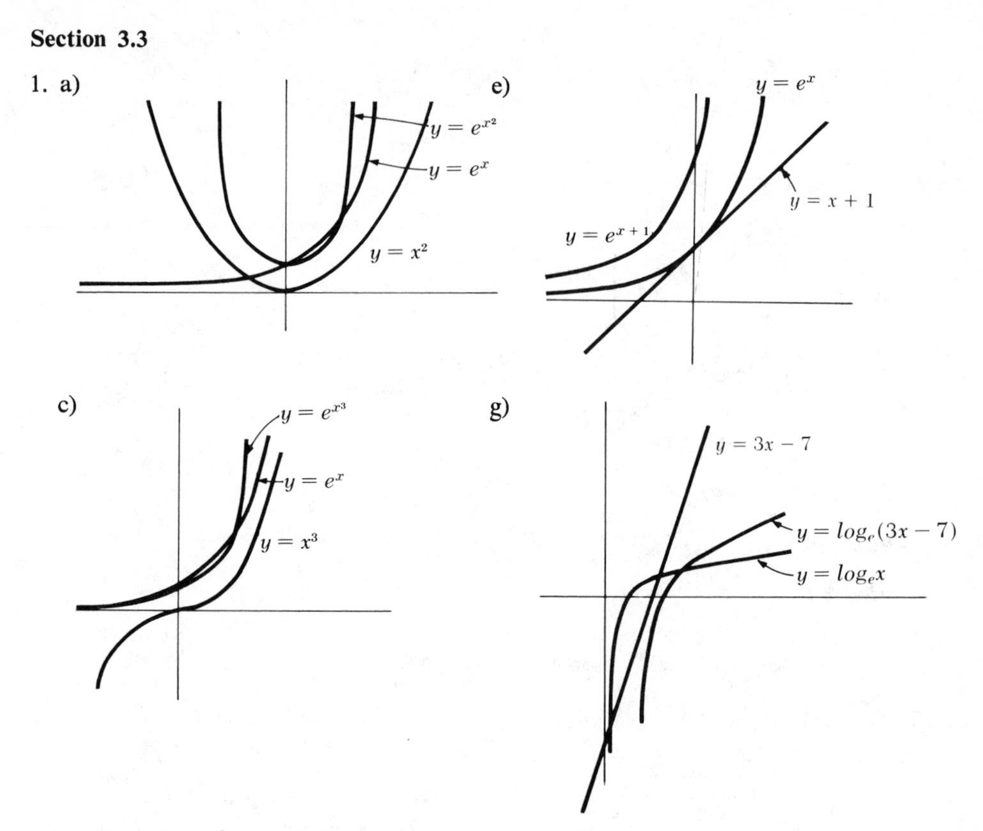

3. Profit on the mth radio is $10 + \log_e (m + 2)$; total profit is

$$100 + (\log_e 3 + \log_e 4 + \ldots + \log_e 12).$$

4. $100e^{0.06}$ dollars.
5. a) $f(0) = 0$; b) $f(x) > 0$ for $x > 0$; c) $f(x) < 0$ for $x < 0$.
6. a) $x > 0$; b) $x = 1$; c) $x > 1$; d) $0 < x < 1$; e) both $\log_e x$ and e^x are increasing as x increases, and hence $e^x \log_e x$ is increasing as x increases.

CHAPTER 4

Section 4.1

1. derivative at	0	2	−1
a)	2	2	2
c)	0	8	−4
e)	−7	−7	−7
g)	not defined, necessary limit fails to exist	$1/2\sqrt{2}$	not defined, $f(-1)$ not defined
i)	0	24	6
k)	2	6	0
m)	−5	−33	9
o)	not defined, necessary limit fails to exist	2	0

2.

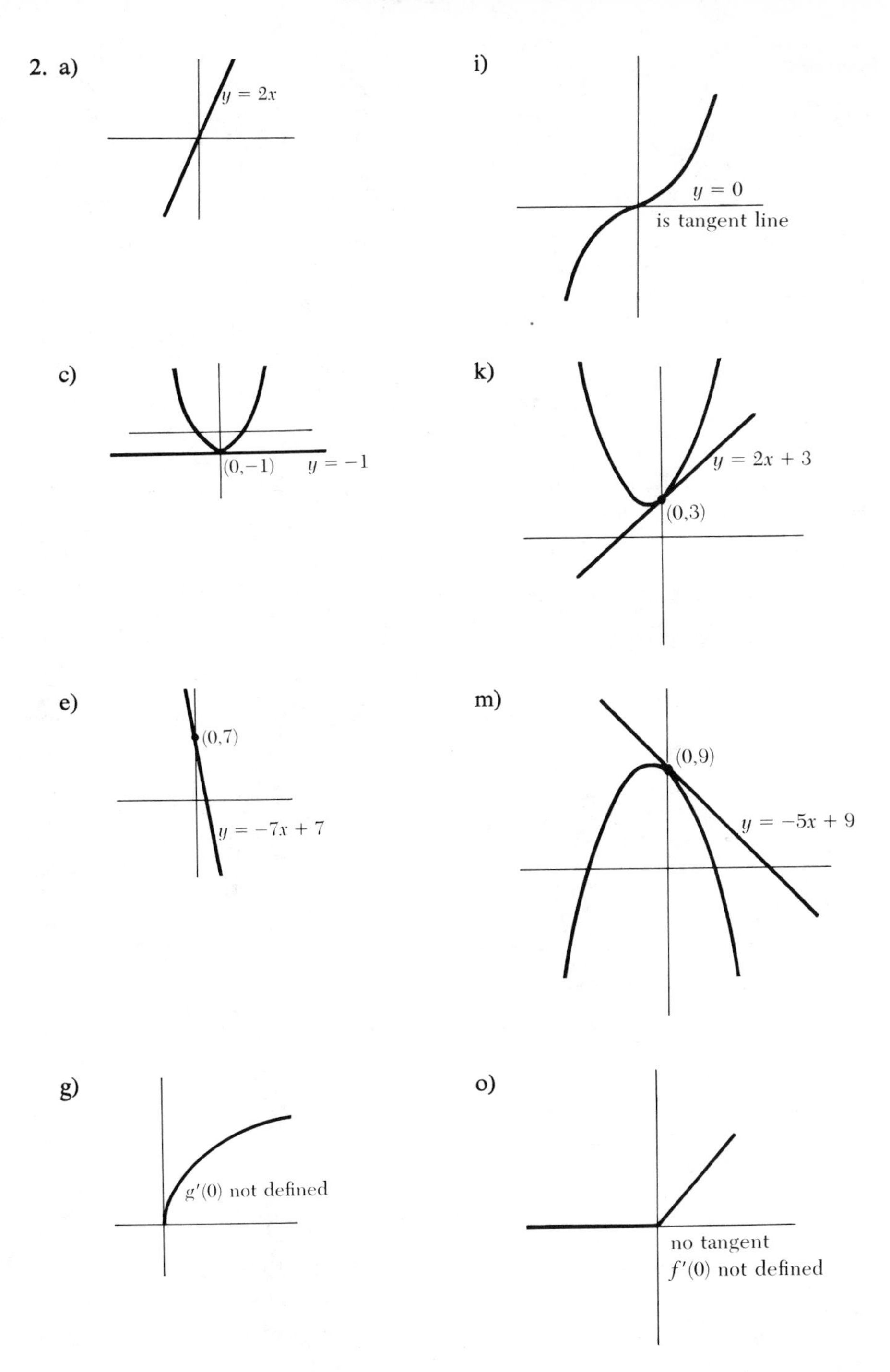

3. a) $v(t) = 160t$; b) 160 mi/min, 800 mi/min, $160(17.5)^2$ mi/min.

6. a) Equation of tangent line to the graph of $f(x) = \sqrt{x}$ at (1, 1) is $y = \frac{1}{2}x + \frac{1}{2}$, with $m = 1/2$ and $b = 1/2$; b) $f(1.01)$ is approximated by 1.005, correct to 4 decimal places $\sqrt{1.01} = 1.0050$.

Section 4.2

1. a) $f'(x) = -5$; c) $f'(x) = 6x$; e) $f'(t) = -1/t^2$; g) $f'(x) = 2x - 3$; i) $f'(x) = 2x + 1/x^2$; k) $f'(x) = -2x/(x^2 + 1)^2$; m) $g'(x) = 3x^2 - 2x$; o) $f'(x) = 0.9x^{-0.1}$, not defined at $x = 0$; q) $g'(x) = 1/x$; s) $f'(t) = -(1/\log_e 3 + \log_3 t)(1/(t \log_3 t)^2)$; u) $g'(t) = 6t + (e^t/t) + e^t \log_e t$.

2.

first derivative	<0	>0	$=0$
a)	all x		
c)	$x < 0$	$x > 0$	$x = 0$
e)	all x		
g)	$x < 3/2$	$x > 3/2$	$x = 3/2$
k)	$x > 0$	$x < 0$	$x = 0$
m)	$0 < x < 2/3$	$x < 0, x > 2/3$	$x = 0, x = 2/3$
q)	$x < 0$	$x > 0$	

3. a)

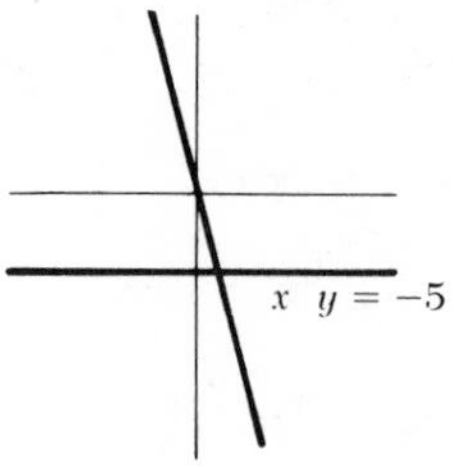

x $y = -5$

e)

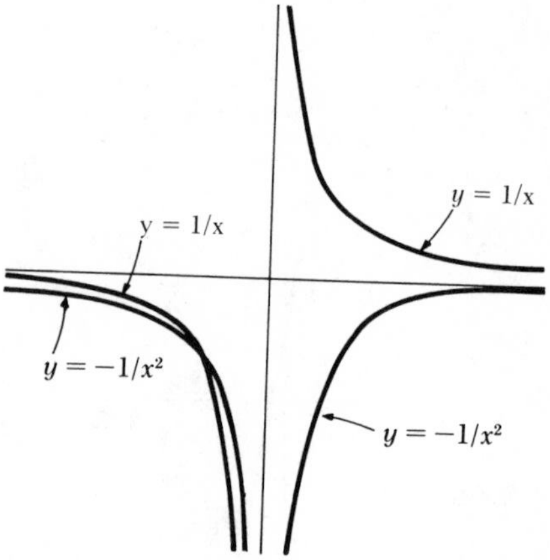

$y = 1/x$

$y = 1/x$

$y = -1/x^2$

$y = -1/x^2$

c)

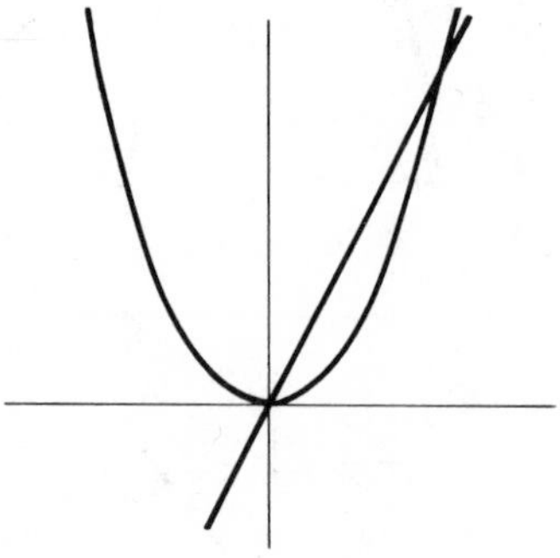

g)

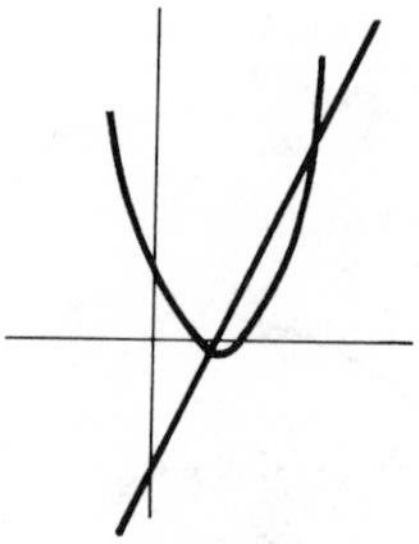

4. a) $f'(x) = 2e^{2x}$; b) $f'(x) = -e^{-x}$; c) $g'(x) = 1$; d) $f'(t) = -3e^{-3t+1}$; e) 0; f) $1/\log_e 3$.

5. $g(x) = f(x)f(x)$ yields $g'(x) = f(x)f'(x) + f'(x)f(x) = 2f(x)f'(x)$; $h'(x) = 3(f(x))^2 f'(x)$; if $g(x) = (f(x))^n$, then $g'(x) = n(f(x))^{n-1}$.

Section 4.3

1. a) $2xe^{x^2}$, composition of x^2 with e^x; c) $\dfrac{(t^2 + 1)e^t - 2te^t}{(t^2 + 1)^2}$; e) $(\log_e 3)3^x$;

g) $(-x/\sqrt{2\pi})e^{-x^2/2}$, composition of $-x^2/2$ with $e^x/\sqrt{2\pi}$;

i) $$\left(\frac{1}{8}\right)\frac{1}{[(((w+1)^{1/2}+1)^{1/2}+1)((w+1)^{1/2}+1)(w+1)]^{1/2}};$$

k) $f'(x)/f(x)$, composition of $f(x)$ with $\log_e x$; m) $(2t+e^t)/(t^2+e^t)$;
o) $((x+1)(x-1))$; q) x^3.

2.

	positive	negative	0
a)	$x > 0$	$x < 0$	$x = 0$
c)	$x \neq 1$		
e)	all real numbers		
g)	$x < 0$	$x > 0$	$x = 0$
q)	$x > 0$	$x < 0$	$x = 0$

3. a) g) c) q) e)

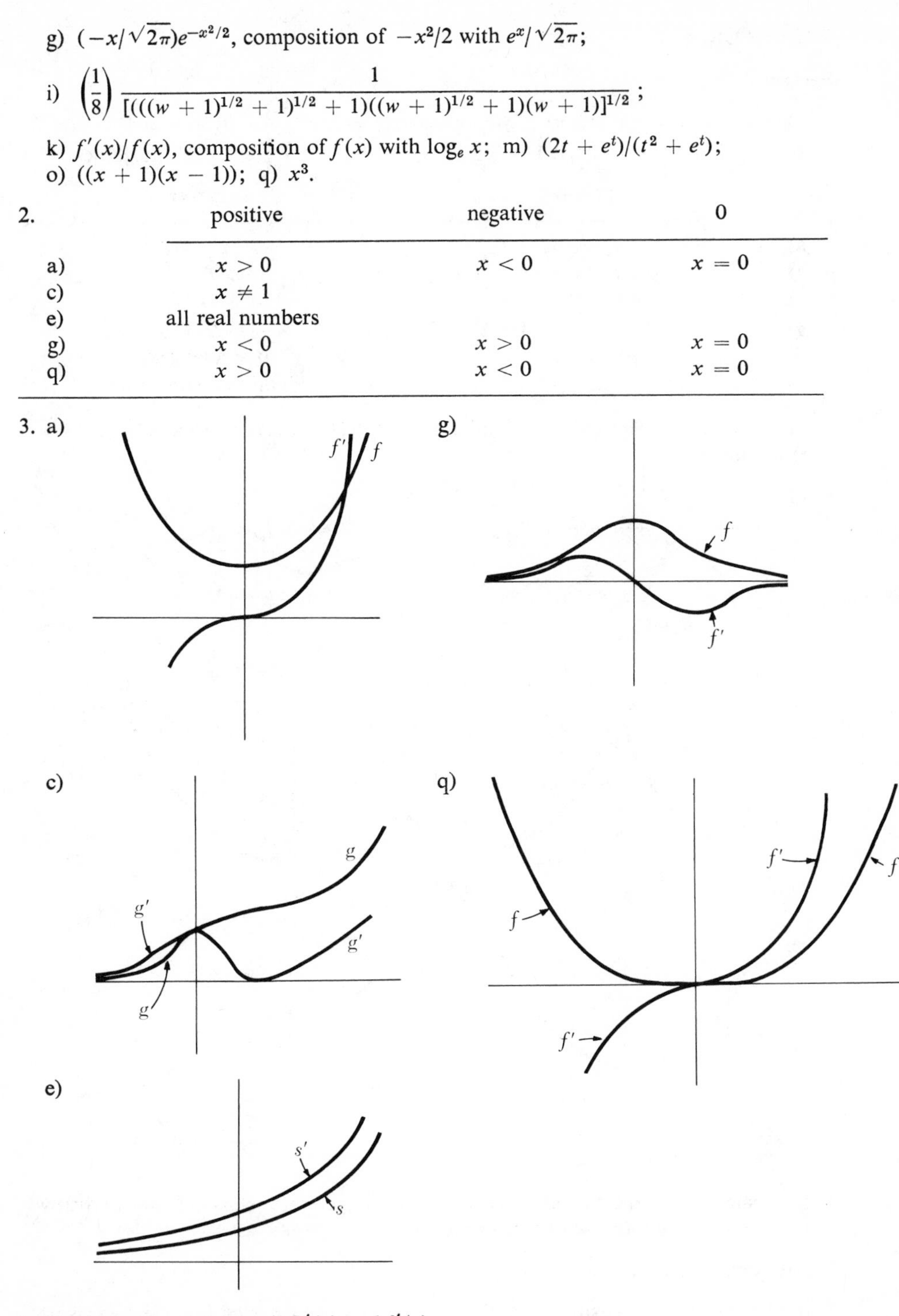

4. a) 1/4 feet per minute, $(kf)'(x) = kf'(x)$.

5. $f'(x) = m'(n(p(x)))n'(p(x))p'(x)$.

CHAPTER 5

Section 5.1

1. a) $-2x - 1$, $x = -1/2$; c) $-2/t^3$, no zeroes; e) $2xe^{x^2}$, $x = 0$; g) $-2t^2e^{-t^2} + e^{-t^2}$, $t = 1/\sqrt{2}$, $t = -1/\sqrt{2}$; i) $1 + \log_e x$, $x = 1/e$; k) $-e^x/(e^x + 1)^2$, no zeroes.

2.

	positive	negative
a)	$x < -1/2$	$x > -1/2$
c)	$t < 0$	$t > 0$
e)	$x > 0$	$x < 0$
g)	$-1/\sqrt{2} < t < 1/\sqrt{2}$	$t < -1/\sqrt{2}$, $t > 1/\sqrt{2}$
i)	$x > 1/e$	$0 < x < 1/e$
k)		all real numbers

3. a) maximum; e) minimum; g) minimum at $-1/\sqrt{2}$, maximum at $1/\sqrt{2}$; i) minimum.

4.

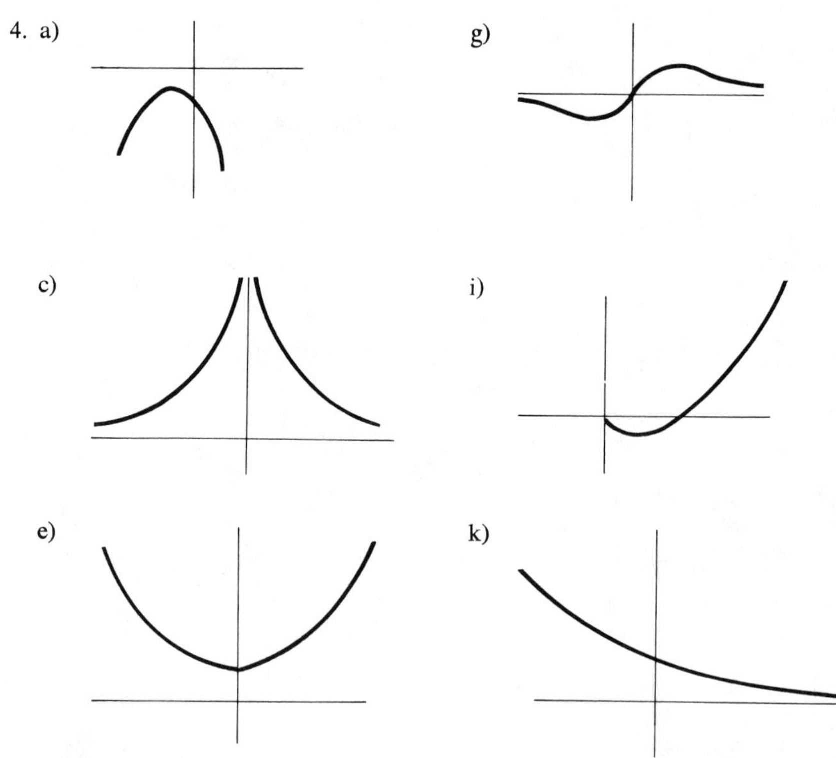

5. Company C has the fastest increasing rate of increase; slope of first derivative measures rate of increase of rate of increase.

7. $f'(0)$ does not exist.

8. minimum value 0, maximum value 1; neither can be detected because $g'(x)$ is not defined in an open interval containing 0 or 1, because g is not defined on an open interval containing 0 or 1.

Section 5.2

1 and 2.

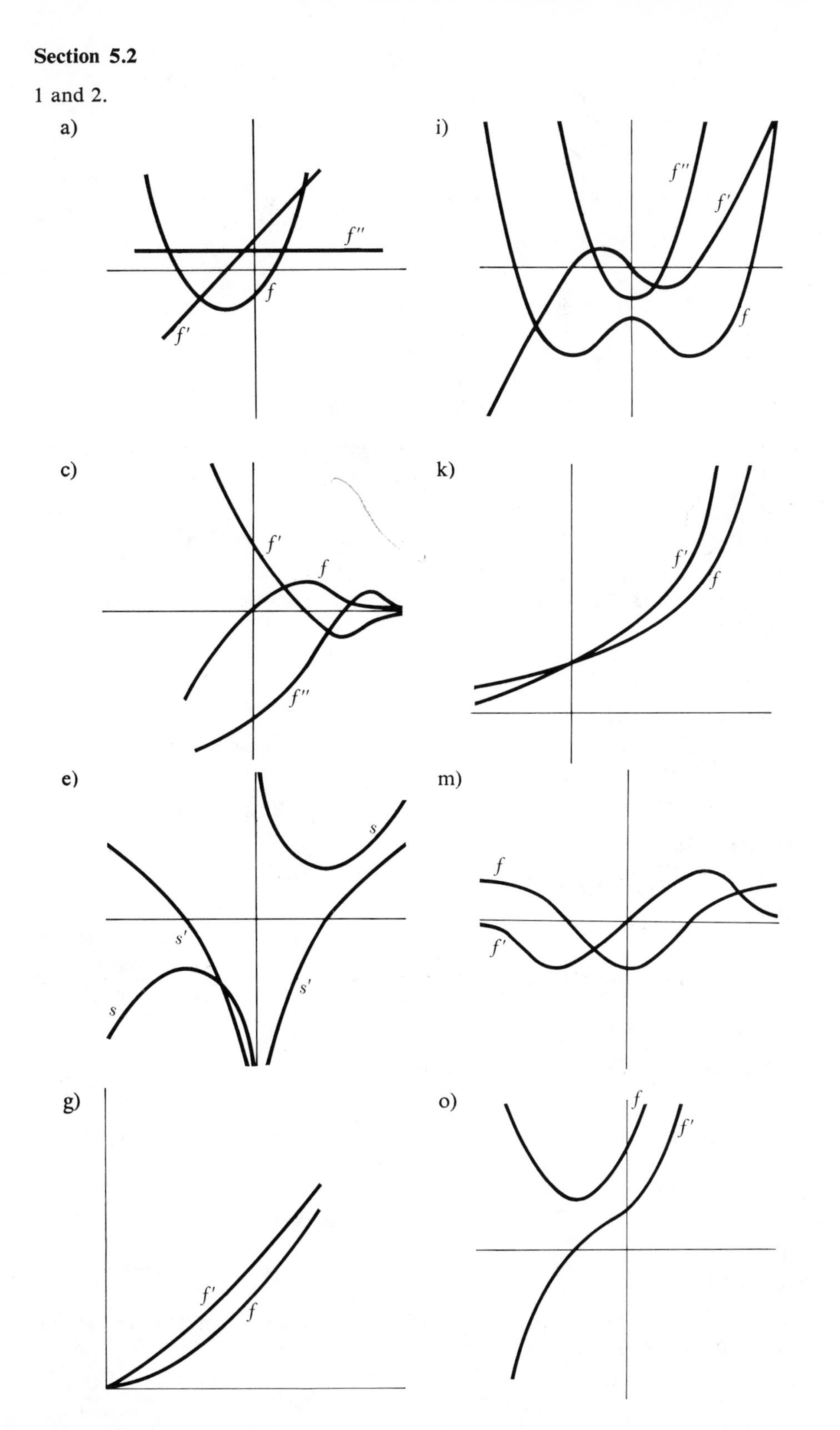

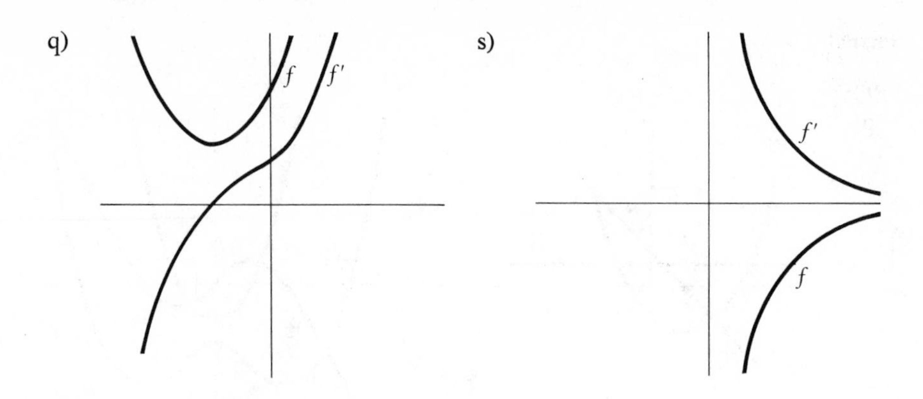

3. Cf. Figures 5–15 and 5–16 for configurations to look for.

4. a) minimum at -3, maximum at 3; b) minimum at 0, maximum at 1; c) no maximum (not continuous at $t = 0$), minima at 2 and -2.

6. Suppose that x and x' are in (a, b) with $x < x'$. Then there is a number c with $x < c < x'$ such that $f(x') - f(x) = f'(c)(x - x') = 0$, which implies $f(x') = f(x)$.

7. Since $(f - g)'(x) = 0$ for all x in (a, b), $f - g$ is a constant function, or $f(x) - g(x) = k$, where k is a constant for all x in (a, b). Therefore, $f(x) = g(x) + k$ for all x in (a, b).

Section 5.3

1. 10, 10.
2. $\sqrt{20}$ and $\sqrt{20}$.
4. 6 feet per second.
8. $\sqrt{18}$, $\sqrt{18}$.
9. Approximately $1045/1.02 \doteq 1024$.
12. Pick right away.
13. Bottom of window is $2/\pi$ feet, height of rectangular portion is $\frac{1}{2}\left(5 - \frac{2 + \pi}{\pi}\right)$ feet.
14. Bottom dimension $2^{-1/3}$, height $2^{2/3}$.
15. $5\sqrt{5}$ feet is the length of the shortest ladder.
16. 1/40 feet per minute.
18. $n = 70$.
19. $x = \sqrt{56}/5$.
20. 4 per cent.

Section 5.4

1. a) $1 + x + x^2/2$; b) approximation 0.99005 is $e^{-0.01}$ correct to five decimal places; c) 1.105, 1.1, 1.052.

d)

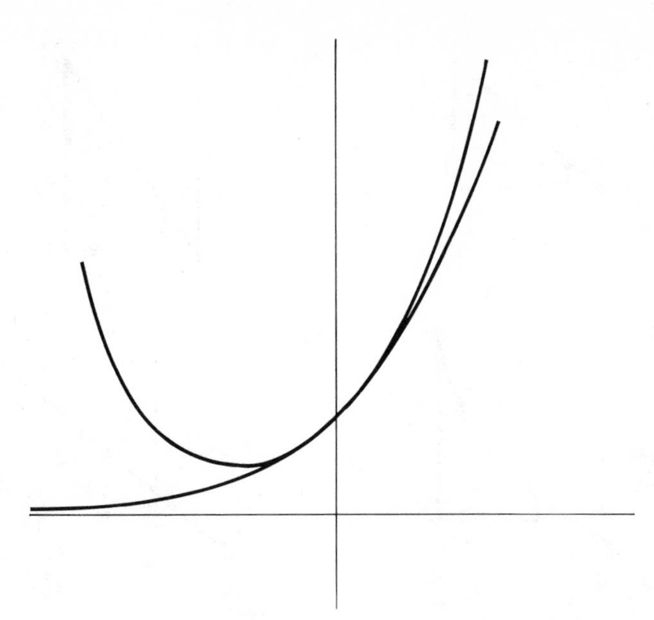

2. a) k; c) $x^2/2 + k$; e) $x^2/2 - 3x + k$; g) $e^x + k$; i) $(5/6)x^{5/6} + k$; k) $e^{x^3} + k$.

3. a) $x'(t) = k/x(t)$; b) $x(t) = k(x'(t))^2$; c) $t = k(x''(t))$, $t = kv'(t)$, $t^2/2 = kv(t) = kx'(t) + c$, $t^3/3 = kx(t) + ct + c'$; d) $t = k(x'(t) + x''(t))$.

5. b) $g(x) = x^3/6 + x^2/2 + x + 1$.

6. $b = 2$, $a = 1/4$, $A = -1/64$, $B = 1/4$, $C = 2$; $f(4.1) = 2.025$, $g(4.1) = 2.0248$; $\sqrt{4.1}$ correct to five decimal places is 2.02485.

CHAPTER 6

Section 6.1

1. a) 5; b) $4\frac{5}{8}$; e) 1961/512.

2. a)

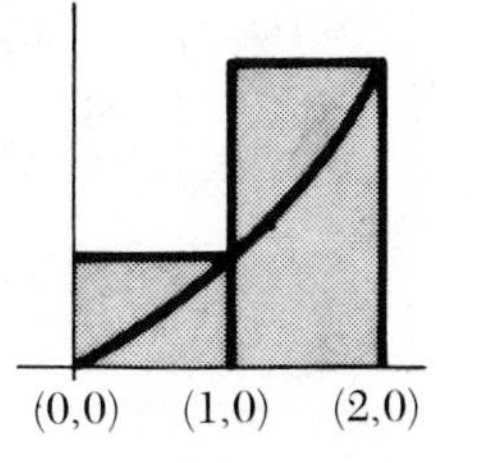

c)

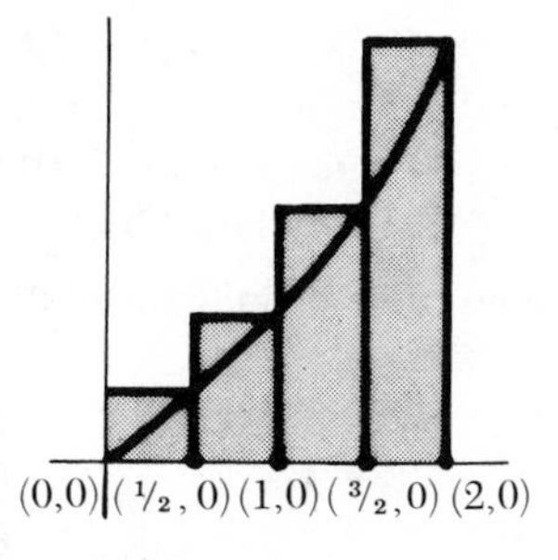

3. $F(x) = x^3/3$, exact area is $2\frac{2}{3}$.

4. a) $(n + 1)/n$, $(n + 2)/n, \ldots,$ $(n + (n - 1))/n$, $(n + n)/n = 2n/n = 2$; b) sum reduces to $(1/n^2)((n + 1) + (n + 2) + \ldots + (n + (n - 1)) + (n + n)) = n(3n + 1)/2n^2 = (3n + 1)/2n$; c) 3/2.

5. a) $0, 1/n, 2/n, \ldots, 4n/n$; c) 20.

6. a) $11\frac{1}{2}$ e) 156 c) 37 g) 32

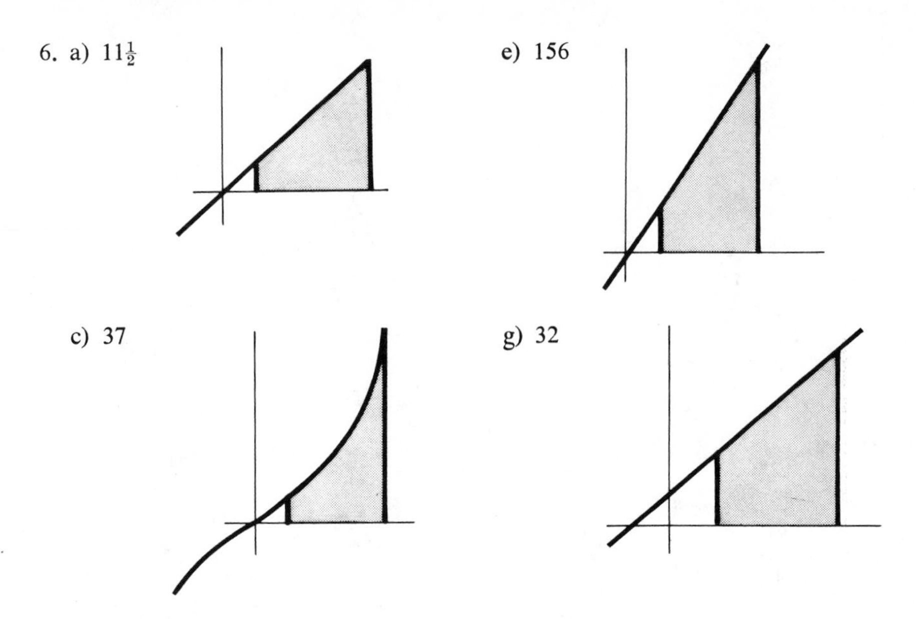

Section 6.2

1. a) $x^3/3 + x + c$; b) $x^3/3 + e^x + x + c$; c) $1/3 + e$; d) $e - e^{-1}$; e) $3/(2\log_e 2)$; f) $\log_e (x + (x^2 - 7)^{1/2}) + c$; g) $x^{1.1} - x^{0.9}/0.9 + c$; h) $(x^2 + 1)^2 + c$; i) $\frac{1}{4}(x^2 + 1)^2 + c$; j) $3 \log_e x + c$; k) $\log_e (x + 1) + c$; l) $x - \log_e x + c$; m) $(1/16) \log_e \left(\frac{x - 4}{x + 4}\right) + c$; o) $x(x^2 - 9/4)^{1/2} - (9/8) \log_e (x + (x^2 - 9/4)^{1/2}) + c$; q) $10 + (9/8) \log_e (6.5) - 3(45/4)^{1/2} + (9/8) \log_e (3 + (45/4)^{1/2})$.
2. a) $(2/3)(3^{3/2} - 2^{3/2})$; b) $5/2$; c) $19/3$.
3. a) $x_i - x_{i-1} \geq 0$; b) area lies below x-axis; c) positive area equals negative area.
4. b) limit is $-\int_a^b f(x)\,dx$.

Section 6.3

1. a) $\frac{1}{4}(x^2 - 1)^2 + c$; b) $\frac{1}{6}(x^3 + 1)^2 + c$; c) 15/12; d) $e^e - e^{e-1}$; e) $x - \log_e x + c$; g) $\frac{1}{2}[(x^2/2)(x^4 + 9)^{1/2} + (9/2) \log_e (x^2 + (x^4 + 9)^{1/2}] + c$; i) $\frac{1}{2}(\log_e x)^2 + c$; j) $\frac{1}{4}(\log_e x)^2 + c$; k) $\log_e (5/4)$; l) $x \log_e x - x + c$; n) $e - 2$; o) $x^2/2 - x + \log_e (x + 1) + c$; q) $\log_e (e^x + 1) + c$; s) $x^2/2 - x + c$.
2. a) area given by definite integral approximates sum in question.

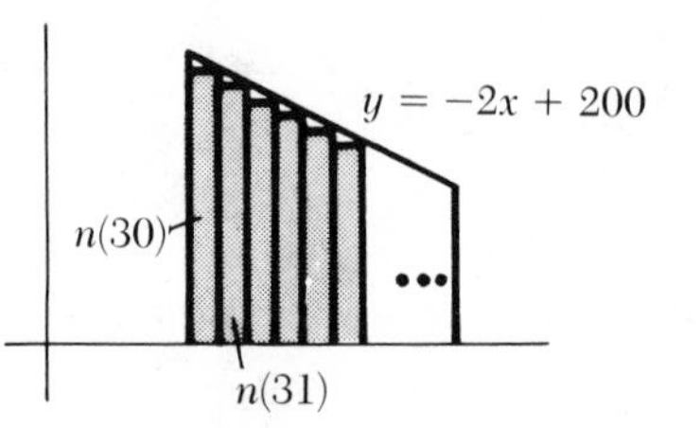

b) 1500.

Section 6.4

1. a) 100 feet/second; b) $10t$; c) $5t^2$; d) 2000 feet; e) 10 seconds; f) 10 seconds.

2. a) $1 - e^{-2}$; b) $\log_e 2$.

3. $\log_e (e^3 + 1) - \log_e (e + 1) = \log_e (e^2 - e + 1)$.

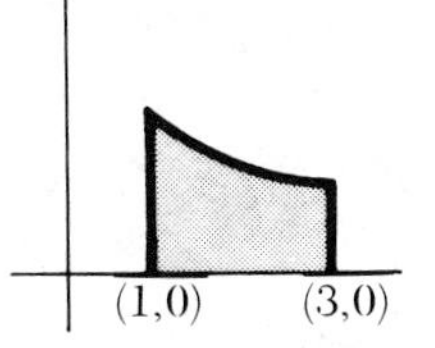

4. 994.9.

5. $250\sqrt{2}/3$.

6. 0.99.

7. a) 96 feet/second; b) 256 feet; c) $50\sqrt{2}/4$ seconds.

9. a) $\log_e 1000$; c) 333,333.

CHAPTER 7

Section 7.1

1. a) 0; b) 0; c) 0; d) 0; e) 0; f) 0; g) does not exist; h) 1; i) 1; j) does not exist; k) 0; l) does not exist.

2.

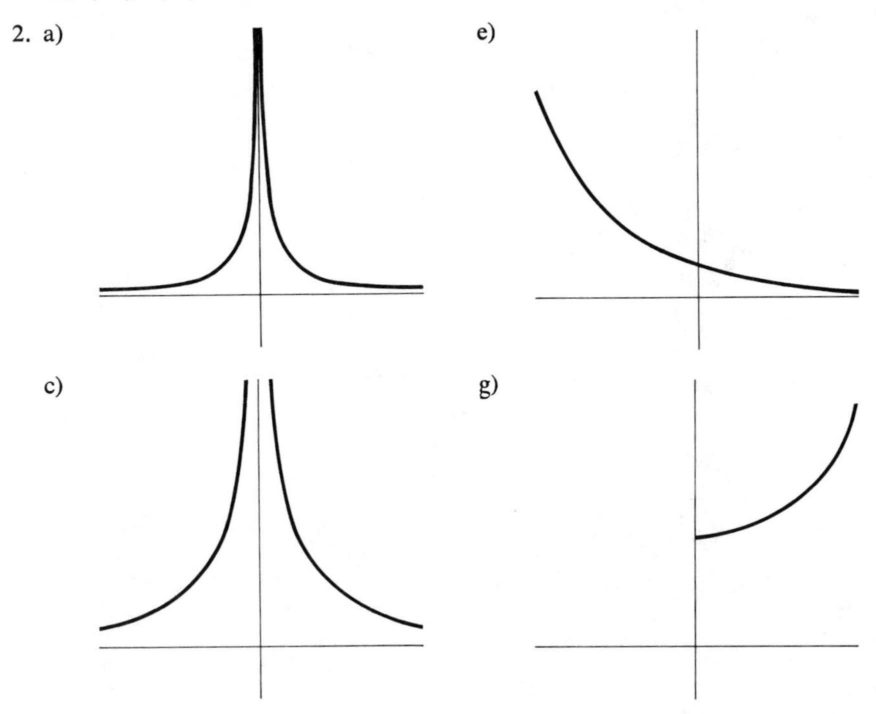

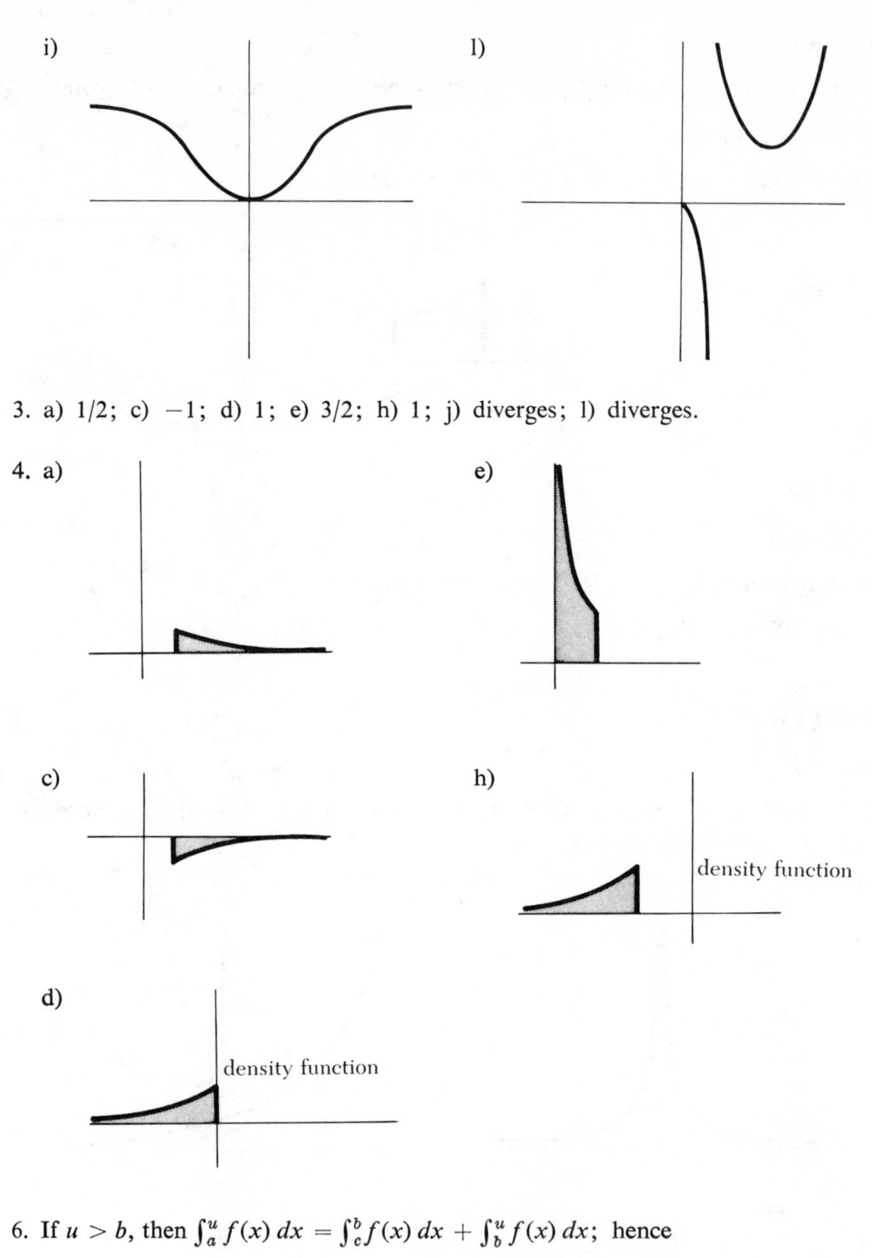

3. a) 1/2; c) −1; d) 1; e) 3/2; h) 1; j) diverges; l) diverges.

6. If $u > b$, then $\int_a^u f(x)\,dx = \int_c^b f(x)\,dx + \int_b^u f(x)\,dx$; hence

$$\lim_{u\to\infty}\int_a^u f(x)\,dx = \int_a^\infty f(x)\,dx$$

$$= \int_a^b f(x)\,dx + \lim_{u\to\infty}\int_b^u f(x)\,dx$$

$$= \int_a^b f(x)\,dx + \int_b^\infty f(x)\,dx.$$

Section 7.2

1, 2, and 3. Answers in the following table.

	Expression (13)	TR	SR	Actual
a)	3/2	2	11/6	2
b)	638/840 $\doteq$ 3/4	1170/1680 $\doteq$ 0.697	1742/2520 $\doteq$ 0.696	$\log_e 2 \doteq 0.693$
d)	$\frac{1}{2}(3.69) \doteq 1.85$	2.42	2.36	2.29
f)	$\dfrac{161{,}341}{346{,}450}$	$\left(\dfrac{1}{8}\right)\dfrac{1{,}256{,}085}{346{,}450}$	$\left(\dfrac{1}{12}\right)\dfrac{1{,}880{,}565}{346{,}450}$	0.458
g)	10.93	25.55/2	38.27/3	13.21
i)	1.39	0.89	0.84	0.89
j)	0.38	0.38	0.38	

CHAPTER 8

Section 8.1

1. a) $f_x(x, y) = y$, $f_y(x, y) = x$; b) $f_x(x, y) = 2xy$, $f_y(x, y) = x^2$; c) $g_x(x, y, z) = y + z$, $g_y(x, y, z) = x + z$, $g_z(x, y, z) = y + x$; e) $S_x(x, y) = x(x^2 + y^2)^{-1/2}$, $S_y(x, y) = y(x^2 + y^2)^{-1/2}$; g) $f_x(x, y, z) = -y^2e^{-xy^2} + 2/x$, $f_y(x, y, z) = -2xye^{-xy^2}$, $f_z(x, y, z) = 3/z$; i) $f_{x_1}(x_1, x_2, x_3) = 16(x_1 + x_2^2 + x_3^3)^{15}$, $f_{x_2}(x_1, x_2, x_3) = 16(x_1 + x_2^2 + x_3^3)^{15}(2x_2)$, $f_{x_3}(x_1, x_2, x_3) = 16(x_1 + x_2^2 + x_3^3)^{15}(3x_3^2)$; k) $g_w(w, z) = e^{z+1}w^{e-1}$, $g_z(w, z) = w^e e^z$.

2. a) $f_x(x, y) = y^2$, $f_y(x, y) = 2yx$, $f_{xx}(x, y) = 0$, $f_{yy}(x, y) = 2x$, $f_{xy}(x, y) = f_{yx}(x, y) = 2y$; c) $h_x(x, y) = 2x$, $h_y(x, y) = -3y^2$, $h_{xx}(x, y) = 2$, $h_{yy}(x, y) = -6y$, $h_{xy}(x, y) = h_{yx}(x, y) = 0$; e) $E_x(x, y) = ye^{xy}$, $E_y(x, y) = xe^{xy}$, $E_{xx}(x, y) = y^2e^{xy}$, $E_{yy}(x, y) = x^2e^{xy}$, $E_{xy}(x, y) = E_{yx}(x, y) = xye^{xy} + e^{xy}$.

3.

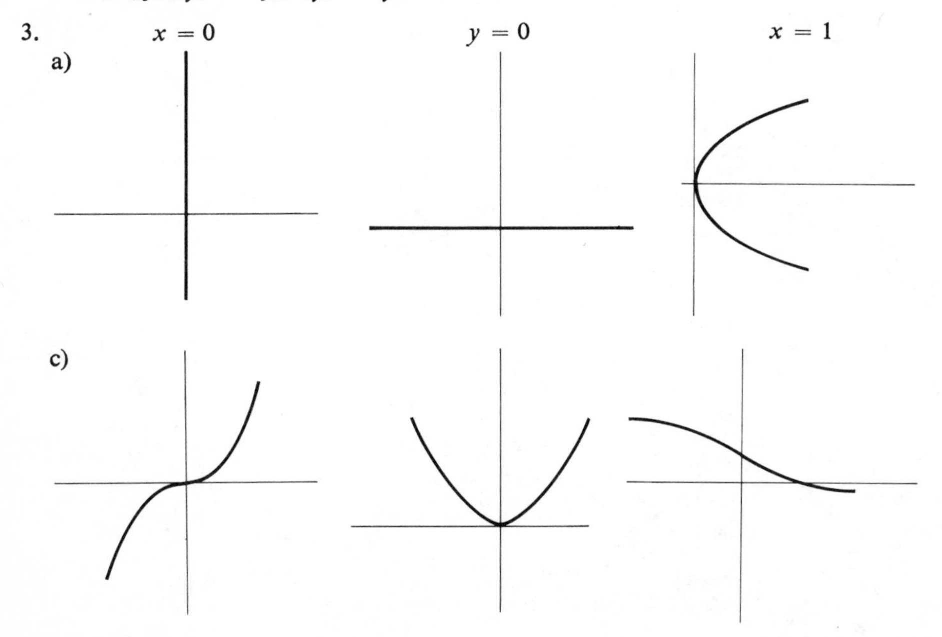

4. a) $T(x, y, z) = 2xy + 2xz + 2yz$; c) $C(x, y) = x + y + 1.875xy$.

Section 8.2

1. a) $(0, 0)$; b) $(0, 0)$; c) none; e) $\{(0, 0, z) \mid z \neq 0\}$; g) $(0, 0, 0)$; i) $(-3/2, -3/2, -3/2)$.

2. a) no information since expression (15) is equal to 0.

5. $x = y = z = 25/3$

7. length $=$ width $= 150\sqrt{2}/18$, height $= 150\sqrt{2}/28$.

Index

Index of Symbols